I0483584

The Nature, Origin, and Profound Implications of Irreversibility

– the Driving Force Behind the Second Law of Thermodynamics

by Ray Bonn
and Lesa Sorenson

(aka Fred Vaughan)

The Nature, Origin, and Profound Implications of Irreversibility
Copyright 2015 Vaughan Publishing
ISBN-13: 978-1516935796
Createspace createspace.com/title/5682502
last updated 5/30/16

For more information from.
or to make comments to,
the authors please email:
fred@vaughan.cc

Seattle, U.S.A.

Acknowledgements

One of the authors has previously acknowledged indebtedness to his family for having tolerated the very lifestyle that makes such endeavors as this possible. It goes without saying that when one spends so much of their time dealing with abstractions much necessary practical work goes undone or is done haphazardly or, worse yet, thanklessly by others. The virtue, if there be any, of such avoidance of usual responsibilities depends, of course, on the ultimate worth of the final product. Some have called this chance of justification as 'moral luck'. Those who pay a heavy price for whatever the worth of the product have little leverage in affecting the assessment of that value that is subjective at best. But having gone down this path again, that author must once again own up to having shirked what others do so graciously.

It is worth considering this dynamic that has given us much of what is grand in our civilization as well as much that is disgraceful. Notable examples are those with much greater talent than that of that current author but who nonetheless largely ignored the significance of the contribution of those who have tolerated their choices. So let us for a moment at least acknowledge the debt we owe the Marie-Anne Lavoisers, Mileva Einsteins, Nora Joyces, Mette Gauguins, Debora Miltons, Jenny Marxes, Cornelia Andersons, and an unaccountably long list of other generous souls who have paid a great price in giving up, or having had taken from them, their own legitimate claims to a full measure of happiness and/or fame.

We hereby acknowledge and thank them all.

Table of Contents

Part II

Part III

Part IV   223

Appendix

Bibliography

List of Figures

Prologue

How this volume took on new life one night on a national TV show in prime time is described in the fictional trilogy *Not Julie*. That was the night the two authors identified on the cover of this volume first met and were ineluctably drawn into collaboration on an effort that would finally explain the origin of irreversibility according to that apocryphal tale. They both agreed that the origin had to be at the submicroscopic level of our existence; it could be no other way. This tome is therefore (somewhat facetiously) put forward as having originated as the result of the interchange that occurred that night as a part of a discussion concerning one of the essays in another of one of the two authors' books, namely in *The Aberrations of Relativity*. When asked about that essay by the other current author, he had pulled an early draft of his investigations into irreversibility from his bag. The other of these two authors immediately grabbed that draft, proclaiming flamboyantly its relevance to ongoing investigations that she too had been conducting into intimately related areas of thermodynamics. The discussion is said to have become quite heated.

According to that story a collaborative effort was suggested and thereafter a publisher produced a contract for the two authors to sign in agreeing to complete this volume. Supposedly both signed although in differing capacities. Subsequently the nature of this collaboration became an even more contentious affair in its own right. Would the work essentially be an updated draft of what the one author had shown the other, with the other collaborator merely writing an extensive forward, or would they bring to bear the products of both their fertile minds in the spirit of a true collaboration? That tension, according to this same fictional account, played out somewhat later in dramatic events at old Yankee Stadium and continued in private communications until finally an amicable resolution was reached in which both authors agreed to co-authorship responsibility. Acknowledging the extreme relevance of the work of both authors had finally seemed reasonable to both and full collaboration ensued.

Of course such work could not have begun nor progressed significantly on prime time television. The ultimate accomplishment required laborious independent effort, coordinated as would always be necessary with such collaboration. That is how science gets done. Years of solitary thought precede more notable conclusions. Personal flare and fantasy have little to do with science and this book certainly bears little resemblance to the flashing and daring of that vibrant conversation allegedly witnessed by so many. But that does not mean that the study of thermodynamics cannot be exciting.

The authors feel that the problem of the mystery of an 'emergent' irreversibility only in our everyday world has finally been adequately refuted by their work. A proper accounting of its true origin at the lowest levels of our existence has been firmly established. Behavior at this low level, beneath the more continuous flow of properties of macroscopic matter where the no-free-lunch and arrow-of-time maxims pertain, has long been acknowledged as contrasting starkly with what is witnessed at the everyday level of our existence. This is because of the supposed pristine character of reversible processes at the lowest levels. Nonetheless, the shroud of irreversibility has now been lifted. It has been demonstrated to make its entry at the submicroscopic level.

At that level of our reality, the physical laws that govern the kinematics and interactions of indivisible particles have been thought to completely abide time-reversible behavior in both classical and in relativistic quantum theories. Until now that pristine view of subterranean realities has been thought to fully and adequately describe all the submicroscopic states by most researchers in the field, although even they readily acknowledge that quantum relativistic corrections are required to earlier descriptions. It has, however, generally been accepted as an outstanding obligation upon anyone who insists on the origin of irreversibility in this idyllic realm to explain this unsuspected and unbefitting aspect of an idealized world as though it were original sin in and of itself with no legitimate excuse in this primordial Garden of Eden. But where on earth could this defiling irreversibility have come from otherwise?

Certainly there is still work to be done, but the mystery of where it comes from and how it arises is now fairly clear. Of course the precise nature of *irreversibility* that is associated with the *second law of thermodynamics* is not all that well understood by a broad audience beyond some vague feel for the fact that there is 'no free lunch' and that there is indeed an 'arrow of time', whatever those pithy phrases might entail. A fairly extensive background into concepts and terminology of thermodynamics is therefore in order. This volume will inform the reader of historical developments that have given us the relevant notion of *entropy* and its ineluctable increases as well as early efforts to define and quantify heat and its effectual relationship to mechanical work.

The authors are convinced of the extreme significance of their efforts. Achieving this level of understanding of the nature of irreversibility is, of course, significant. But just as discovering the source of the Nile did not change its flow one iota, discovering the origin of the 'arrow of time' does not alter the flight of that arrow in any way. It certainly does not justify renewed searches for a fountain of youth or perpetual motion machines. But hopefully it will help the reader better understand exactly why such endeavors are doomed to inevitable failure.

Truth be told as science demands, both authors are products of a silly fiction, their combined names are but a pseudonym for R. F. Vaughan who hopes that all this obfuscation will not, as it should not, detract from the significance the reader will assign to his work. For as a philosophy professor once proclaimed to a student who had raised the tangential issue of a particular philosopher having, according to that student, been a 'horrible man' for allowing his family to live in poverty while he promoted his philosophy, and therefore why would anyone take heed to anything he had written: "The truth of a proposition has nothing to do with who may have stated it." That is a truth this author wishes to strongly reinforce by having attributed his ideas to fictional characters.

R. F. Vaughan
2015

Forward

"A certain professor of thermodynamics was known to give the same final exam every year, always consisting of just the single question: 'What is entropy?' One day an assistant suggested that it might be better to ask a different question now and then, so the students wouldn't know in advance what they would be asked. The professor said not to worry. 'It's always the same question, but every year I change the answer.'"[1]

In addition to confusion with regard to exactly what entropy *is*, current scientific explanations of the associated irreversibility and the ineluctable increases in entropy are complicated, unsatisfactory, and completely incorrect. This problem is so impenetrable in fact that in over two centuries of notable attempts by the greatest scientific minds there has still been no explanation that is credible. The ubiquitous increases in entropy seem, however, to only affect the happenings at the macroscopic level of our everyday existence for which no process is completely reversible. Processes that are irreversible like those we witness every day with the naked eye are *ipso facto* those for which entropy is increased. But there has seemed to be no origin of this dire trend at the submicroscopic level where the answers to virtually all of the difficult problems of physics have been resolved.

The asymmetry between past and future in dealing with most of the processes we encounter in our everyday lives is readily apparent. That asymmetry is associated with what is referred to as 'irreversibility' that has seemed to be totally absent in all of the underlying submicroscopic interactions that are correlated with these very same processes.[2] The unidirectionality of these 'aging' processes at the macroscopic level of our existence has given rise to the term 'the arrow of time'. What is referred to as the *second law of thermodynamics* specifically addresses the asymmetries associated with irreversibility of natural phenomena. A parameter called *entropy* was introduced as a means of quantifying the facts associated with this aspect of thermodynamics – facts that are more familiarly acknowledged under the rubric of there being 'no free lunch'.

---

[1] *http://www.mathpages.com/home/index.htm*

[2] There is, of course, an isolated example of kaon decay that occurs at the microscopic level acknowledged as displaying such an asymmetry as well, but that is certainly not involved in the phenomena we will discuss. However, the situation with kaon decay is not that different from what is involved in what we will discuss in detail.

A 'reversible' process is one that could proceed equally well either forward or backward with respect to clock time. An irreversible process is one that could only proceed forward in time; the physical laws of nature would be violated if it were to proceed in the reversed temporal direction. This distinction seems fairly self-explanatory since we can witness many physical processes that are (at least in principle) reversible. These include the progression of the planets in their orbits, the swing of a pendulum, or hockey pucks colliding on a frictionless sheet of ice. Films played in reverse of orbiting planets, swinging pendulums, or colliding hockey pucks would appear physically realistic to us because proceeding in reverse is, in fact, equally realistic. The dynamics would satisfy Newton's laws of motion when played in either direction because Newton's laws are themselves time-symmetric.

Processes that seem obviously 'irreversible' also come readily to mind: Expansion of smoke from a fire or gas from an explosion into the open air, molecules released into a room from an opened deodorizer bottle, an infant maturing into an adult, and the progression of a terminal disease. In contradistinction to films of reversible processes, a film of smoke, expanding gas escaping a bottle, a time-lapse film of a child growing up, or the progression of a terminal disease would make no sense to us if run in reverse. Wisps of gas being sucked back out of a room and forcing their way into a bottle would certainly not correlate with anything that would make any physical sense to us. That is, of course, because it does not proceed according to anything we have come to understand about nature. A diver proceeding in reverse back up to a diving board would defy logic, as would children growing smaller, the dead coming back to life, etc.. It is inconceivable that the water molecules that were forced away upon a diver's entry to the water could somehow converge to force him back up into the air.

However, this issue of irreversibility is not quite that simple. Orbiting objects will eventually decay into lower orbits, pendulums eventually stop, and hockey pucks will ultimately come to rest with videos of these situations played in reverse making no sense whatsoever. Also, if we consider the situation of an expanding gas more carefully we must acknowledge that the gas consists of many individual molecules, each with some particular velocity; they collide with each other much as would the hockey pucks mentioned above. Each of these collisions is envisioned to obey Newton's laws of motion, which are, as we mentioned above, time-reversible, with no friction involved. Thus everything envisioned as being involved in the escape of gas from a bottle or a diver submerging into the water is perceived as time reversible. So the entire process of expansion into and throughout a room, or a swan dive into water would seem also to allow reversibility. This is in that same sense of not specifically violating any of Newton's laws of mechanical motion if it were to occur. And yet... we know that it *cannot* happen.

It is a predominant view of those who should know better that such reversible behavior is merely precluded by extreme improbability, i. e., that there is no physical reason why a diver could not dive into a placid pool and be summarily ejected. They believe there to be a finite (however small) probability that the diver could complete his dive only to find himself once again standing on the diving board, i. e., that irreversibility is simply a matter of probability. The authors (as probably most of our readers will) find that explanation absurd.

Whatever the explanation of irreversibility happen to be, 'the second law' is the aspect of thermodynamics that addresses the nature of irreversibility. It addresses the issue of ineluctable dissipation of thermal energy available to perform mechanical work. It declares that with the passage of time an equilibrium condition of uniform temperature with a stable distribution of energy among constituent submicroscopic particles and maximum entropy of the system will be reached. There are many seemingly different but equivalent statements of this law. One is that heat will always flow away from the higher and into to the lower temperature substance and never in the opposite direction. Another time-honored statement is that no cyclic process can be established to completely convert heat into mechanical work. All these statements make sense as descriptions but they do not sufficiently explain the associated situation or certainly the cause of irreversibility.

The thermodynamic state variable, entropy has been defined to formalize such statements quantitatively, thereby allowing a rephrasing of the law to state succinctly that entropy of an isolated system cannot be decreased. A process in which entropy increases is *ipso facto* irreversible so we at least have some sort of mathematical handle on the concept of irreversibility. The second law of thermodynamics embraces this expression of fact and is the only basic law of physics to exhibit asymmetry with respect to time. So naturally this statement is tantamount to *the law of irreversibility*. It is argued by some, and for good reason, that the second law of thermodynamics actually assigns a meaning to the concept of time as we know it.

But how the concept of irreversibility is captured by the notion of entropy and what exactly *is* entropy are questions we must still address. We proceed by accepting that irreversibility is intimately connected to the meaning of entropy. And we embrace entropy as breaching the gap between the macroscopic and lower level submicroscopic aspects of this notion of irreversibility in thermodynamic systems.

There are two separate contexts of definitions of entropy. It was originally defined at the macroscopic level as a state variable useful in the context of classical thermodynamics. It is also defined in more direct association with the kinetic theory of molecules of gases in particular at the submicroscopic level as well as statistical mechanics at a somewhat higher level. But those are descriptions *not* explanations.

The first law of thermodynamics states that energy cannot be created nor destroyed. Therefore, the total amount of energy in any closed system from one point in time to the next must remain constant. But when thermal energy performs work, that energy is said to have been 'expended', i. e., used up. The extent to which energy can be used (in this sense of being 'used up') to produce a mechanical change in a system or its environment depends on the extent to which the available thermal energy is unevenly distributed throughout the system. When a system is in complete equilibrium, no 'useful' work can be done by that system even though there may yet be a large amount of total energy vested in that very system.

The second law of thermodynamics tells us that energy always tends to become more uniformly distributed throughout the system, or at least more 'stably' distributed in accordance with the Maxwell-Boltzmann distribution of particulate energies, thus restricting the amount of useful work it can do. Entropy provides the measure of that limitation associated with uniformity (i.e., stability) of the energy distribution. In the submicroscopic definition it addresses the number of equivalent combinations of particles with given energies in its current energy distribution at the submicroscopic level. This number of equivalent distributions increases as systems tend toward equilibrium situations even though the total energy of the system remains the same.

But why would that be of such profound significance?

It has become quite acceptable in scientific circles to presume that irreversibility is due to what is denominated 'emergent' phenomena, i. e., that it just pops out of nowhere when systems become sufficiently complex. This presupposes that when natural phenomena become so complicated that we can no longer comprehend all the intimate details, the associated causes of what we observe need not, therefore, derive from within the detailed behavior of what we observe. But that rather glib and irrational answer defies centuries of scientific tradition.

In 1986 John Earman stated in the abstract to his scientific paper entitled 'The Problem of Irreversibility' that "*After reviewing recent literature from physics and philosophy, it is concluded that we are still far from having a satisfying explanation of the nature and origins of irreversibility. It is proposed that the most fruitful approach to this problem is to concentrate on conditions needed for a rigorous derivation of the Boltzmann equation.*"[3] That assessment has remained relevant and his recommendation is directly in line with the one that is taken by the current authors, although much more than elastic molecular collisions must be taken into account.

Other explanations have been attempted to be sure, but largely without success. The accepted Statistical Mechanics explanation seems to be that although there is always a tiny possibility of entropy decreasing, the probability is so small as to justify belief that it *cannot* happen. In other words, the vagaries of the motions of molecules in the air in a room could all happen to be such that at some instant all the molecules would end up in one corner of the room. We are advised that such violations of the Second Law of Thermodynamics, while remotely possible, will almost certainly never happen so that the law is, so to speak, 'good enough for government work'. But that explanation defies logic. It can easily be shown that the conservation of momentum makes it provably impossible for the center of mass of all the molecules of air in a closed room to convene to a corner of that room.

Another major attempt at explanation has been to interpret the trend of increasing entropy in accordance with certain versions of the standard cosmological model that implies that the universe must necessarily be winding down or at least trending in a non-reversible direction. This latter explanation of the origin of irreversibility apparently seems appropriate to some because irreversible behavior is ubiquitous throughout space-time in all epochs and aspects of the universe up through life processes. They, therefore, argue that this is all a part of the 'big picture' of our universe at the *highest* level, demanding a top-down rather than a traditional bottom-up reductionist explanation. They seem to feel that such an explanation satisfies the requirement that whatever the cause, it is universal at the highest levels even if not present at the *lower* levels of the universe. This argument denies an origin, or even a necessary presence, of irreversibility in submicroscopic phenomena of which the entire universe is comprised.

Of course the current 'concordance' model of cosmology suggests an overall cooling of the universe itself with no stable stopping off point short of an absolute zero temperature. Such a description of our

---

[3] *Proceedings of the Biennial Meeting of the Philosophy of Science Association* © 1986, Vol. 2, pp. 226 – 233.

universe certainly does admit to an asymmetry between past, present, and future with an associated 'arrow of time' but it sheds no light whatsoever on the specifics of the supposed origin or cause of these phenomena in our 'down to earth' everyday lives.

The identification of a specific cause' of irreversibility associated directly with this cosmological perspective was proposed by those who adhere to what is called the 'astrophysical trend'. It is compatible with the universe having itself had a dramatic, although still unsatisfactorily explained, origin rather than as being characterized by a stationary state. This view of something out of nothing is, of course, a consensus view enshrined in the currently accepted concordance version of the standard model of cosmology, but 'something for nothing' certainly does not accord well with the 'no free lunch' maxim of the second law of thermodynamics to which it would be applied in this case.

This anticipation of a top-down explanation of irreversibility is to be contrasted to the opinions of those of us who suspect that, although so far undiscovered and unaccounted, irreversibility does occur (and in fact originates) in interactions at the submicroscopic level of our existence. The specific aspect of the phenomenon that does account for this behavior has been there all along waiting to be discovered. The authors are convinced that they have indeed uncovered it – and it was not hiding.

A protracted debate with a friend concerning this issue of origins is what precipitated one authors' intense interest in this aspect of nature. To allow a shroud of mysticism concerning the origin of irreversibility to render illegitimate a reductionist approach to science that has been inherently successful as a primary methodology for reducing observed phenomena to causal explanations at lower levels of reality seemed totally nonsensical to him. The approach of modeling behavior at a lower level to explain higher level phenomena has benefitted all of the physical sciences including most-notably, a majority of the phenomena grouped under the general heading of thermodynamics.

Even the possibility that there could be an *acausal* and irrational *emergence* of irreversibility between levels of description of reality would challenge the most time honored of scientific methodologies. It is akin to accepting a magical incantation for *creatio ex nihilo* as a modus operandi for science. Welcoming the creation of 'something from nothing' should be particularly anathema in a discipline that prides itself on the realization that there can be 'no free lunch'.

But this irrationality doesn't stop or rather didn't start there. For if the motion and collisions of on the order of Avogadro's huge number of submicroscopic particles in one mole of a substance is to be accepted as the complete explanation of 'heat' as is so often averred, for example, how then can there be no such quantity associated with what occurs in the collision of two such entities in relative motion? This is particularly telling when it is precisely those motions and collisions, with no direct reference to radiation or anything else that are said to collectively constitute what heat *is*. Some vestigial aspect must be present there. At what level of complexity are we to believe that *this* quality called 'heat' makes its clandestine entry if not at the level of the motions and collisions themselves to which it is attributed? With regard to the somewhat analogous attribution of the explanation of 'pressure', for example, the situation is quite different for the simple reason that between the concepts of a particle exerting a force at a point and a force being exerted over an extended area there is no such categorical incompatibility. And since reductionism is employed in these disciplines, why would it be invalid with regard to irreversibility.

The friend of one author subscribed to the 'astrophysical trend' that has contributed a modernized creationist view of thermodynamics to the conversation. Photons generated in an instant during the hottest early microseconds of our universe are supposed by some advocates to have more or less been 'sucked out' into a vast chill of some obscurely conceived space-time in which they suppose our (and perhaps other) universe(s) to be imbedded or obscurely comprised. Although current cosmologists (when limiting themselves to our observable universe) more typically conceive space-time itself as doing the expanding 'beneath' the photons rather than as constituting an empty receptacle into which they proceed. The effects, certainly with regard to a cosmological redshift, would be very similar. This process according to the astrophysical trend is supposed to have begun at a time when our universe was more interesting to current cosmologists, when if they are correct, it began as a miraculous bang – a *big* bang mind you.

At any rate, it is fashionable in physics nowadays to consider the big bang with its recently incorporated accomplice, the 'God particle', as the origin of everything otherwise unaccountable by traditional physical concepts. According to this notion of such an origin of the universe, including in some versions its 'universal' physical constants, an associated origin for irreversibility must seem only reasonable. In this regard the *astrophysical trend* was perhaps a trendsetter.

But any explanation of irreversibility derivative to an origin of the universe would require links to explain why increases in entropy apply at the local level on a time-scale for which any evolution of the universe would be irrelevant. This criticism certainly recommends itself when one considers the notable fact that the avowed expansion involves separation speeds that are considerably less than 10^{-20} cm/sec for separations of up to a few meters in a gas for which velocities are upwards of 10^5 cm/sec even at room temperatures. It is difficult to envision how any such effect could possibly contribute substantially to phenomena we witness here, now. This is *the* problem, not just an irrelevant corollary of the problem. The other way around is physics on its ear. If an astrophysical trend is the cause of irreversibility, then what on earth is the mechanism by which its effect is produced? This 'answer' leaves the most pertinent questions unanswered.

This same friend was not convinced by the author's arguments that the causes of irreversibility must occur at the submicroscopic levels of reality. The arguments had not given him the slightest pause to even reconsider his position or attempt refutation of his own hypothesis. In an exchange the friend opined merely that: "The evidence for arrows is so overwhelming that I don't know where to begin, and there's little point in boring you with elaborate lists. A simple intuitive example that springs to mind is the surface temperatures of the Sun, which range from an inner one over a million degrees, whereas the outermost 'layer' is maybe a few thousand degrees." *

Facts concerning the vast variations in temperature and density throughout the various domains in our universe were, of course, not news to the author. But when succinctly asked as conclusion, "*How could such a steep gradient be possible without cold outer space?*" the author realized what was at issue between this friend and himself – the scope of the philosophical dilemma with which they wrestled.

These age-old problems of philosophy will never go away; conjectures that attempt to resolve associated issues will only cause these truly meaningful problems to be reformulated with successively more relevance accruing down through the ages, but forever nagging at our heels nonetheless. With a renewed understanding of the nature of the gulf between them, i. e., opposing convictions that irreversibility originates at the *bottom* or at the *top* of this tiered structure we refer to as reality, that author proceeded to attack the horns of the Parmenidean dilemma – Heraclitus's river that is always the same yet always different, a universe that seems always to be expanding and cooling and yet is arguably always the same.

In this volume the authors will explain their current understanding of how irreversible changes of characteristic aspects of our local universe can persist even in a continuously stable universe. There need be no trend toward universal expansion and cooling with or

---

* Private e-mail communication dated Sun, 17 Nov 2002.

without eventual collapse associated with this thermodynamic trend. The universe is not a grandfather clock in need of some grandfatherly figure or 'God particle' to either start or rewind it. That is merely the myth of our forefathers revisited.

Certainly, to accurately assess whether a universal trend exists one must sample suspected behavior over substantial segments of time and space with samples that can be justified as representative of the phenomena for which the trend is presumed to apply. To this end one must have a valid model of the behavior of the entire system being sampled. Of course when the system under test is the entire universe one can run into unique modeling problems to say nothing of notable errors in categorical reasoning.

In Heraclitus's metaphor of the river one must include much more than the solidity of its banks, the fluid that flows between those banks, and the gravitational gradient that forces it to the sea if one is to resolve the paradox of identity in flux. And this *is* the paradox we must address. More is required than the addition of mountains, foothills, and valleys through which tributaries flow into the river, and more than models of the occurrence of seasonal rain and snowfall if one is not to eventually have ones modeled river run dry or fill the modeled seas to overflowing. The evaporation of water cycles it back to the river's source. One cannot forget to model the part that closes the cycle. One must complete the loop in any valid model if one is ever to have a chance to understand an equilibrium situation. Without complete models of all such logical loops, a lasting equilibrium will always seem to be impossible. It is no different with the *astrophysical trend* to which we have just focused some attention.

We have concentrated briefly on the suspect top-down universal trend explanation of irreversibility because the associated cosmological facts do in fact bear on the issue at hand. Although as we will show, far from being the cause or origin of irreversibility, the appearance of this trend *results from* the cause of irreversibility at the submicroscopic level of our reality rather than the other way around. There are in fact irreversible interactions that take place between submicroscopic particles. These interactions between molecules in a gas involve, in addition to direct collisions, exchanges of photons of radiation and these exchanges are major aspects of thermodynamic systems at the submicroscopic level. They are in fact the association with the thermal 'blackbody' radiation. These photons are affected by second order relativistic Doppler caused by the relative motion of the interacting particles. Irreversibility arises in these interactions because both radiation and mechanical energies are involved in the same transaction between particles. The fact is that the relationship between formulas for the conservation of energy and momentum for photons of radiation differ substantially from the relationship for particulate matter. This singular fact results in the irreversibility of viable interactions between particles mediated by photons. So irreversible interactions do, in fact, occur even at the submicroscopic level, which, as it turns out, is not so pristine after all.

Certainly trends, i.e., gradients and change, are essential to our nontrivial world, but that does not preclude stability – whether cosmological or other. Certainly there are gradients of temperature in the universe just as there are gradients associated with the flow of water through river banks.* But that does not in itself suggest either that rivers will all one day run dry nor that the oceans will overflow in a material analogy of Olbers' paradox. There is more subtlety in heaven and earth than that! Olbers had not the slightest conception of magnitudes involved in either the separations in space or finite lifetimes of stars or he would not have conjectured as he had, and others would not have wasted so much time on his supposed paradox. It should come as no surprise that open loop models of rivers have them running dry, and open loop models of thermodynamic systems can never adequately represent the ineluctable ebb and flow of entropy.

In resolving irreversibility at the submicroscopic level it has been necessary to augment Boltzmann's kinetic theory beyond two types of interaction and to more fully elaborate necessary constraints on the emission and absorption of radiation in Einstein's quantum theory of radiation. It is in the interactions between these domains where irreversibility enters. It has been incumbent upon us to close major loops left open by the scope of their analyses. Boltzmann could not have foreseen the impact of mediated interactions involving quantized photons, nor certainly relativistic effects.

A comprehensive model has had to be developed to incorporate complimentary mechanical and radiational aspects of a thermodynamic system. The mediated interactions between molecules that do not involve direct collisions always reduce the relative velocity of the interacting molecules, which is very *entropic* behavior. In this way, individual submicroscopic processes 'use up' otherwise useful energy and increase entropy even at the submicroscopic level.

Yet another form of interaction involving both radiational and particulate dynamics is the scattering of radiation by arrays of charges

---

* Nor, of course, should anyone presume that the authors do not accept an only superficially similar trend of global warming as a reality. It obviously is.

within a thermodynamic system. 'Forward' scattering in particular has traditionally been considered to involve conservative forces that do not alter the energetics of either the ensemble of particles or the radiation field. We show that this too is an over simplification whose correction has profound consequences of irreversible behavior, producing what have been considered 'cosmological' effects. The major loops that must be closed in this regard involve the origin of the ubiquitous hydrogenous plasma with 24% helium by weight and the supposed disappearance of mass (and information) in black holes. There is increasing evidence that black holes do indeed errupt spewing forth hydrogenous plasma to produce the 24% helium in generating the gamma radiation that after prolonged redshifting caused by irreversible scattering becomes the microwave background radiation. And as we will show, its temperature does not reflect the kinetic temperature of the particulate matter by which it is scattered.

In order for a closed loop model of the universe rather than an obscurely defined 'trend' to apply to the universe, the universe itself must exist in a 'stationary state', i. e., at the highest level it does not change. We are born and die; stars are born and die. Galaxies are formed and ultimately collapse into black holes. But that is not significant change at the highest level. And through all of this, entropy increases without punctuation. But does it?

No. Black holes swallow all the accumulated entropy increases. All the information (memories) about us and our descriptions of observations of all that exists and has existed will be obliterated – nothing will be left but a quiet black hole with no more information to give. The history of life on earth, our planet, solar system, the Milky Way galaxy – all will be lost along with the entropy that is our history. Entropy will indeed finally decrease to near zero which will defy the Second Law of themodynamics. But a rebirth of beginnings and new information will be heralded by the eruption of that black hole – the ultimate phoenix rising from the fire. So that finally after its demise has been completed, entropy will once again begin the one way journey with which we are so familiar.

And that, if you will, is the story of irreversibility.

Part I

1: Historical Developments in Thermodynamics

"On one hand, thermodynamics is a formal system that allows us to deduce interesting consequences from a few simple laws, wherever those laws apply... But they do not apply everywhere. Thermodynamics would have no meaning if applied to a single atom. To find out whether the laws of thermodynamics apply to a particular physical system, you have to ask whether the laws of thermodynamics can be deduced from what you know about that system. Sometimes they can, sometimes they can't. Thermodynamics itself is never the explanation of anything – you always have to ask why thermodynamics applies to whatever system you are studying, and you do this by deducing the laws of thermodynamics from whatever more fundamental principles happen to be relevant to that system." – S. Weinberg[1]

As the top-most tier of a multilevel hierarchical structure, classical thermodynamics addresses relationships between and among the temperature, pressure, and volume of what is typically a gaseous substance when various processes are employed to determine the amount of physical work performed on or by that system. Entropy provides a measure of the efficiency of performing work as a part of this top level explanation of thermodynamics.

The phenomena observed and described at this level can be given a more fundamental rationale by describing associated phenomena in the submicroscopic domain of molecules, atoms, electrons, photons, etc.. Statistical mechanics provides an intermediate level between the everyday realm of classical thermodynamics and the submicroscopic

---

[1] Steven Weinberg, "Can Science Explain Everything? Anything?" *The New York Review*, May 31, 2001.

quantum realm at which the molecular and atomic interactions occur that result in the observed phenomena. It embraces computational methods from which to calculate thermodynamic parameter values that characterize a system by applying the pertinent statistics to lower level molecular interactions. More recently *information theory* has been built on similar premises having to do with information content. We will not concern ourselves much with intermediate levels of description beyond the awareness of the importance that has been placed upon them.

Thus, the topic of thermodynamics can be variously handled at multiple levels, giving comprehensive results within the restrictions of each domain that are nonetheless applicable in some sense to all the others. And it is important that these levels be, in fact, compatible despite using unique methods pertinent to completely different physical constructs. Without the theoretical and experimental work applicable to each of these descriptive levels, our understanding of thermodynamics would be incomplete. Well… more incomplete than it is. That an inconsistency and incompatibility exists within this overall framework is indeed disconcerting. That will be the focus of later chapters.

We will describe investigations of a host of dedicated researchers notably including those of Robert Boyle (1627–1691) and Jacques Charles (1746–1823) who established relationships between the pressure, temperature, and volume of gases. A consistent description of the behavior of ideal gases resulted. That formed the theoretical basis of thermodynamics. To explicate the reasons behind all the confirmed relationships, Sadi Carnot, Gay-Lusac, James Prescott Joule, Rudolf Clausius, Clerk Maxwell, Josiah Gibbs, Ludwig Boltzmann, William Thomson (Lord Kelvin), J. J. Thompson, and many other pioneers peered under the floorboards of the steam engines and other machines that exploit the parametric relationships of classical thermodynamics. They explained observed relationships in terms of the behavior of astronomical numbers of submicroscopic particles in measurable quantities of substances. That lowest level description of kinematic interactions worked extremely well in explaining the behavior of gases.

Classical thermodynamics is where all of the basic thermodynamic parameters and their relationships were experimentally established. It is a theory of macroscopic (large-scale) systems based in large part on what are called 'ideal gases' in (or near) their equilibrium condition. Strictly speaking, it primarily applies to systems for which a well-defined temperature, pressure, volume, amount of material substance, and an associated quantity of physical 'work' can be unambiguously defined.

historical background

Like so much of science, the Greeks first introduced the concepts that would later be grouped under the heading of 'thermodynamics'. The Greek terms $\theta\varepsilon\rho\mu\eta$, 'therme' meant 'heat' and $\delta\upsilon\nu\alpha\mu\iota\varsigma$, 'dynamis' meant 'power'. The term 'heat' has been associated very directly (although the authors consider the exclusiveness of this accepted association to be somewhat erroneous) with the 'dynamic' movements of the particles at the submicroscopic level. However, firming up exactly what that association is (whether direct or indirect) and what that association suggests from epistemological perspectives has been problematical to say the least. It is here where the inimitable 'irreversibility' of thermodynamics enters the picture. On the other hand, the use of the term 'power' and its quantitative associations with mechanics has been less obscure. The equivalence of thermal and mechanical energy is solidly based.

In classical thermodynamics it is essential that clear definitions of the *system* under test, its boundaries, and its *surroundings* be made explicit. Any such system under consideration will inevitably be composed of innumerable tiny particles in dynamic motion. Although it is to a large extent the statistics of those motions that define its properties, there are typically rigid constraints that all aspects of these constituents of the system remain within the bounds of the system and not stray out into its surroundings. Certain thermodynamic processes involve exchanges across defined boundaries but with the processes constrained in ways that preserve the value of one or another parameter such as, for example, the 'internal heat' of the system being preserved in 'adiabatic processes', etc..

There are, of course, 'equations of state' that succinctly characterize the behavior of thermodynamic systems for which there is no need to consider the motions of constituent microscopic particles. Internal thermodynamic energy and various forms of potential energy can be expressed in ways that are useful for determining conditions for thermal equilibrium and dynamic processes without referring explicitly to lower level constituent particles or processes.

It was Joule who first showed that the work done in manipulating the volume and pressure of a gas could be transferred to and from the thermodynamic energy reflected by the temperature of the gas. He thereby established the principle of the conservation of *total* internal energy and an equivalence relationship between the two subsidiary forms of energy. That then became the basis of the concept of the conservation of total energy, i. e., first law of thermodynamics.

The relationship between the amount of heat that is lost and the amount of work that is performed by a system was as significant as the relationship between mass and energy would become to a later generation. The proportionality constant between mechanical and thermal energy has been given his name. It is claimed that the number he obtained for this constant, which was 722.55, was inscribed on his tombstone. That value is now known to poorly approximate the accepted conversion constant, which is 778 foot-pounds per British thermal unit.

Sadi Carnot would publish his *Reflections on the Motive Power of Fire* in 1824. This was a general discourse on heat, power, and engine efficiency. This marked the beginning of thermodynamics as a modern science, but it was certainly not the *beginning*. William Rankine wrote the first thermodynamics textbook in 1859 at the University of Glasgow. James Joule had previously designated 'thermodynamics' as the science of relations between heat and power. Joule also did pioneering work to demonstrate the powerful concept of conservation of the total combined mechanical and thermal energy. But even earlier in 1849 as previously discussed, the composite term 'thermo-dynamics' had come into play as derived from Greek roots by William Thomson (Lord kelvin) in his paper *An Account of Carnot's Theory of the Motive Power of Heat*.

These texts expand upon what became known as 'the ideal gas law' whose historical development preceded these later documentaries by more than a century. We will now discuss that law briefly.

development of the ideal gas law

The significance of Otto von Guericke's vacuum pump in 1650 is certainly of note with regard to demonstrating the relationship between density of a gas within an enclosed volume and the associated pressure on that volume. Although certainly over dramatized, the air contained in two pressed semi spheres was evacuated to demonstrate that two teams of horses could not then separate them.

Robert Boyle and his assistant Robert Hooke later demonstrated more precisely the nature of the inverse relationship between the pressure and volume and/or density of encapsulated gases. As a graphic presentation, consider the cylinder and piston shown in figure 1.1. In this figure, there is pressure applied to the piston at the right to constrain the volume of the gas contained in the cylinder. As the pressure is increased, the volume of the enclosed gas will decrease. Here we assume that the cylinder is embedded in an environment held at a constant temperature, T.

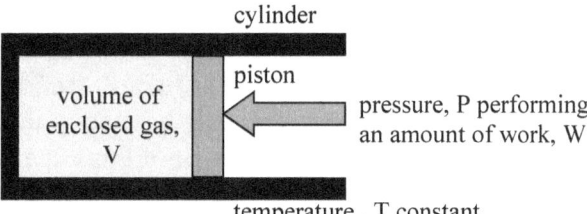

Figure 1.1: Mechanism for assessing Pressure vs Volume relationship

What Robert Boyle initially demonstrated was that the product of the pressure and volume of a gas is a constant in this case. In particular, if V_o and P_o are the initial volume and pressure at temperature T_o, then when the temperature is increased or decreased, the new value of the product will be:

$$P \, V = P_o \, V_o \, (1 + \alpha \, T)$$

where T is the new temperature of the gas and α is a constant that depends on the units used for measuring temperature. This leaves the selection of a scale of temperature measurement so as to eliminate the additional additive constant from the proportionality relationship. The Kelvin temperature scale is thus defined specifically to accommodate this formula. The extrapolated proportionality of products of pressure and volume with temperature depends upon it. At a temperature of *absolute zero* on this Kelvin scale the product vanishes and the expression above becomes $P \, V = T$.

The Kelvin temperature scale relates to the Fahrenheit and Celsius scales as illustrated in the following table:

Kelvin (K)	Celsius	Fahrenheit
0	−273.15	−459.67
233.55	−39.6	0
273.15	0	32
373.15	100	212
1273.15	1000	1800

Of course this direct proportionality pertains exclusively to what are called 'ideal gases', which term is an idealization appropriate to gases whose physical conditions of pressure and temperature preclude their condensation. Although it applies only to this case it is nonetheless extremely useful over a broad range of conditions for most all gases.

The direct proportionality between the volume and temperature of an enclosed gas known as Charles' law was attributed to unpublished work from the 1780s by Jacques Charles. Refer to figure 1.2 in which it is assumed that the pressure, P is held constant throughout changes in the temperature of the gas. But all of this expanded state of knowledge was a direct result of the experimental and theoretical work of a host of individuals in addition to Charles, including Mariotte and Gay-Lussac.

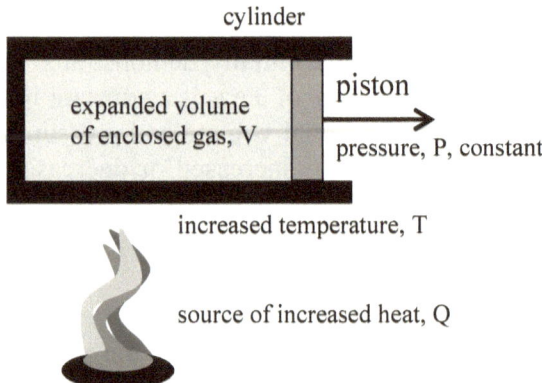

Figure 1.2: Mechanism for assessing Temperature vs Volume relationship

So although Sadi Carnot (1796-1832) may be considered the 'father of thermodynamics', there was a wealth of pertinent research results already available in this general area well before he happened on the scene. From such efforts, the *ideal gas law* was ultimately formulated to characterize the state of a gaseous substance that is in equilibrium. This law is typically expressed with the following formula:

$$P V = N R T$$

The various implied relationships are illustrated in figures 1.3.a through 1.3.f. below. In this ideal gas formula P, V, and T represent pressure, volume, and temperature, the latter being measured on the Kelvin (K) scale. In this equation N represents the amount of the gas present in the designated volume, specifically the number of moles of the gas, under consideration and R is the 'universal gas constant:

$$R = 8.3143 \cdot 10^7 \text{ erg / mol K}$$

A mole of a substance is an amount that contains Avogadro's vast number N_A of molecules or basic units of a substance. This is based on

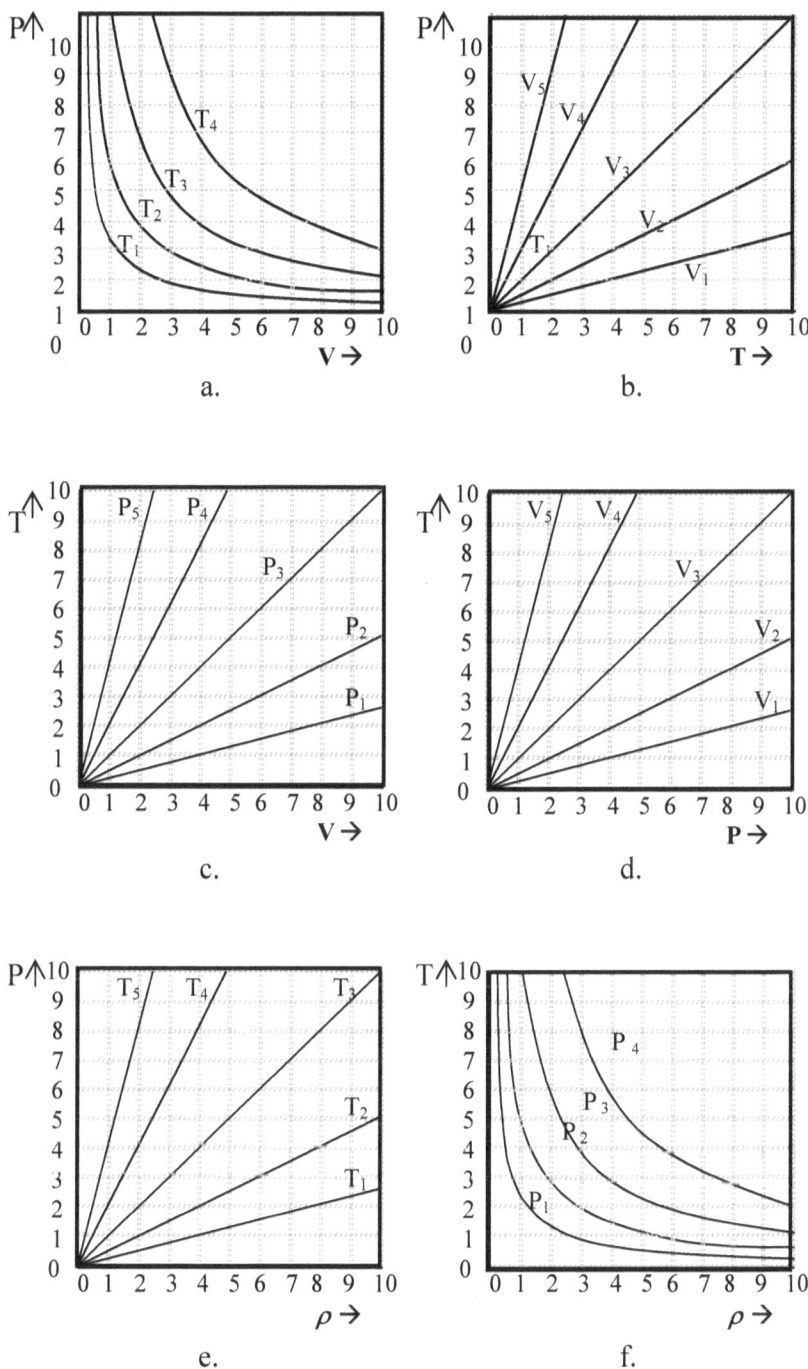

Figure 1.3: Ideal gas law implications

Avogadro's (and a host of other researchers) discoveries that given similar circumstances of temperature and pressure, the same volume of different gases will contain the very same number of molecules even though that amount of gas will exhibit a different total weight that is proportional to the total atomic weight of the individual atoms of the elements chemically bonded into each molecule.

$$N_A = 6.0225 \times 10^{23}$$

This value is based specifically on the number of carbon 12 atoms (periodic table entry 12) contained in 12 grams of carbon 12.

Conversion between moles and molecules is where the macroscopic and submicroscopic descriptions of thermodynamic systems are made to correspond. If the ideal gas constant R is converted to apply to individual molecules in the system, we obtain what is referred to as Boltzmann's constant k, where:

$$k = R / N_A = 1.3805 \cdot 10^{-16} \text{ erg} / \text{K}$$

Graphic depictions of the relationships in this single ideal gas law equation are provided in figures 1.3.a through 1.3.f on the previous page. You will notice that in each of these figures a different one of the basic parameters is held constant.

The final two plots in figure 1.3 describe the associated density changes (i. e., numbers of moles) contained in a constant volume. The density ρ of a gas in a given volume V of substance is:

$$\rho = N / V.$$

the numbered laws of thermodynamics

The second law of thermodynamics – the 'no free lunch' dictum involving irreversibility – is familiar to virtually everyone even though a fuller understanding of that aspect is not well understood at all. The other laws, all of which are easily understood, and even *how many* other laws of thermodynamics there are, remain much more obscure to most people. These 'laws' are the following:

0. The 'zeroth' law of thermodynamics – included as an afterthought which could not preempt the existing numerical order of the other laws although its primacy is obvious – states that thermodynamic equilibrium is an equivalence relation. That is, if two systems are

separately in thermal equilibrium with a third system, they must be in thermal equilibrium with each other as shown in figure 1.4.

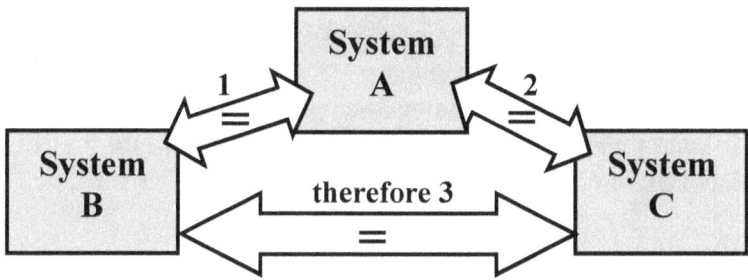

Figure 1.4: Illustration of the zero-th law of thermodynamics

1. The first law of thermodynamics pertains to the conservation of internal energy, U. Namely, the *change* in internal energy ΔU of a closed system is equal to the sum of the amount of heat energy Q supplied to the system minus the amount of mechanical work W done by the system.

$\Delta U = Q - W$

This depends, of course, on the established equivalence relation between mechanical and thermal energy.

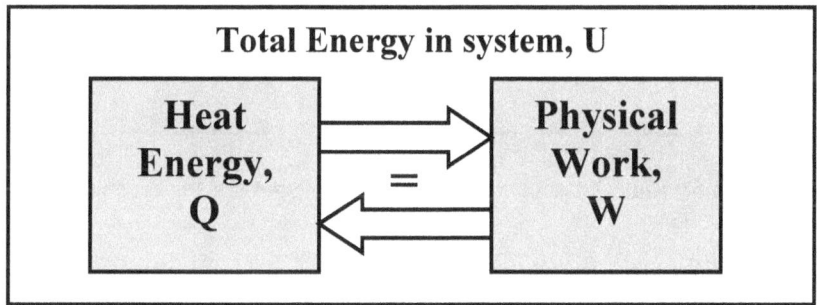

Figure 1.5: Illustration of the first law of thermodynamics

As we will discuss with regard to a reversible mechanical work cycle performed on and by a thermodynamic system, physical work is expressed as:

$W = P \, \Delta V$

Similarly for reversible heat transfers an expression to account for the heat added or removed from a system is provided by:

$$Q = T \, \Delta S$$

where S is defined as the 'entropy' of the system. And yes, entropy is more of a heat-related than a work-related entity.

2. The second law of thermodynamics is specifically about the nature of irreversibility, entropy in particular. The total entropy S of any isolated thermodynamic system can never decrease over time.

$$\Delta S \geq 0$$

It is also a statement of the fact that heat always flows from the hotter to the cooler system. Entropy as it is associated with irreversibility and as a measure of the inefficiency of a system will be discussed in detail further on.

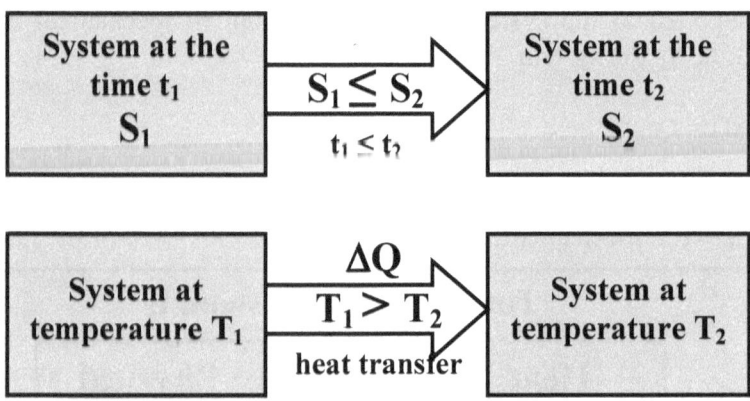

Figure 1.6: Illustration of statements of the second law of thermodynamics

3. The third law of thermodynamics pertains to an absolute zero temperature. The Kelvin scale uses Celsius units. As the temperature of a system asymptotically approaches absolute zero on this scale, all of its dynamic processes virtually cease and the entropy of the system will asymptotically approach a minimum value. In other words, the entropy of a system would be zero at absolute zero Kelvin. This (we will discover) implies that it is impossible to reach the absolute zero temperature by any finite number of processes.

4. Sometimes Onsager's reciprocal relations are referred to as the Fourth Law of Thermodynamics. These express relations between

flows and forces in thermodynamic systems that are *not* in equilibrium, but for which a notion of local equilibrium exists. For example, the heat flow per unit of pressure difference and the mass density flow per unit temperature difference are equal. This is what was shown to be necessary by Lars Onsager using statistical mechanics.

Because the object of this treatise is to obtain a more valid understanding of the nature of irreversibility, we have begun with a methodical approach to understanding all aspects of *reversibility* with regard to the various laws of thermodynamics even though we will provide only cursory discussions of those classical topics. Of course, the nature of the various defined properties of thermodynamic systems must be understood as well as the processes both reversible and irreversible that are understood in terms of them. That must be a primary initial focus of any such investigation. So we will continue with a discussion of these aspects of thermodynamics.

2: Reversibility

"Reversible transformations belong to classical science in the sense that they define the possibility of acting on a system, of controlling it. The 'dynamic object' could be controlled through its initial condition. Similarly, when defined in terms of its reversible transformations, the 'thermodynamic object' may be controlled through its boundary conditions: any system in thermodynamic equilibrium whose temperature, volume, or pressure are 'gradually' changed passes through a series of equilibrium states, and any reversal of the manipulation leads to a return to its initial state. The reversible nature of such change and controlling the object through its boundary conditions are independent processes. In this context irreversibility is 'negative'; it appears in the form of 'uncontrolled' changes that occur as soon as the system eludes control. But inversely, irreversible processes may be considered as the last remnants of spontaneous and intrinsic 'activity' displayed by nature when experimental devices are employed to harness it."
— I. Prigogine & I. Stengers (p. 120)

It is clear that the ideal gas law pertains exclusively to stable conditions at or near equilibrium. It is an equation that describes the *state* of a system when it is not in a situation of flux. However, when changes occur very slowly the system can evolve with the ideal gas law still approximately describing the system through a series of continuous changes. Furthermore, these changes can be reversed. So of course there is no Santa Clause or free lunch Virginia... but sometimes there nearly is... ... if one is patient enough and can perform the necessary processes very slowly.

physical concepts of reversibility

Reversibility is a concept associated with 'conservative' forces that include all of the classical laws of natural, most notably gravitation and electromagnetic forces. Friction on the other hand is not a conservative force and processes involving it are obviously not reversible. Reversibility results when involved forces can be succinctly formulated using ordinary differential equations.

The concept of a 'force' was clarified by Sir Isaac Newton in his second law of motion, i.e., that *for every action there is an equal and opposite reaction*, which he formulated as:

$$F = m\,a$$

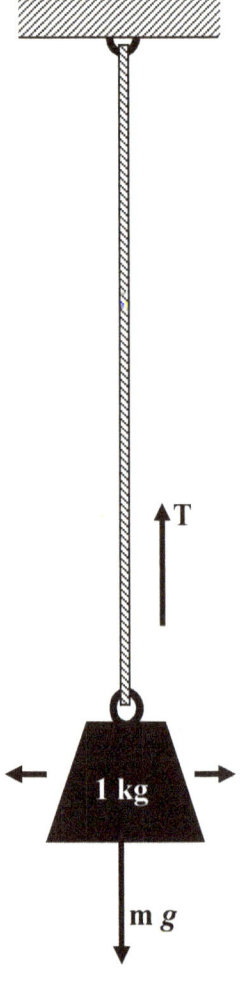

where F is the applied force, m is the mass of the object that is being moved, and a is the acceleration of the object, where a is defined as:

$$a = \frac{d}{dt}v$$

equal to a rate of change of velocity *Rapid change increases the force required to move an object.* According to Newton's laws, moving an object an incremental distance ds, requires energy E, where,

$$E = F\,ds$$

Consider the one kilogram mass hanging on a long rope illustrated in figure 2.1. In this case the one kilogram mass experiences implicit downward acceleration due to gravity that is equal to g = 9.8 m/second^2. Upward tension T on the rope must overcome the tendency of the mass to fall at this increasing rate. Therefore to lift the weight one meter requires a minimum expenditure of energy of:

$$E = 1 \text{ kg x } 9.8 \text{ m sec}^{-2} \text{ x } 1 \text{ m} = 9.8 \text{ joules}$$

This is a minimum since by 'hurrying the effort' *more* acceleration (and therefore force) is involved.

In contrast the amount of energy required to slowly move the mass one meter to either side, for which no force of gravity need be overcome and the

Figure 2.1: The application of Sir Issac Newton's laws of motion

acceleration can be kept to near zero, requires virtually no force or energy to be expended at all.

reversibility in mechanical systems

We first address mechanical applications of reversible processes for which we will later show thermodynamic analogies. We have chosen to implement an elevator that will require a minium expenditure of energy. In such applications we speak of 'reversible' changes for which the system is never more than 'differentially' out of balance.

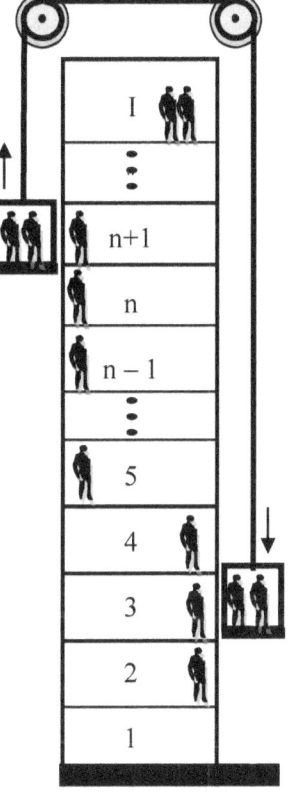

In such idealized cases changes in the state of the system can be reversed to establish a more or less perpetuated cycle to be exploited in performing a meaningful amount of *work* in the everyday sense of that term. If such changes are implemented very slowly in infinitesimal steps, the work performed will achieve the ultimate efficiency. Admittedly this cannot exploit the total energy invested in such a system, and some will have to be added, but still, the best that can be realized.

a pulley-driven elevator

Pullies and cables support the design of an elevator as shown in figure 2.2 that would not require a motor to drive it. In this case a person going up would have to be exchanged for equal weight going down with someone getting onto and off opposite platforms for each ascent/descent of the elevator as shown.

Figure 2.2: Reversible pulley-driven mechanical elevator mechanism

To begin operations a person will have to get onto one of the elevator platforms and another person doing the same thing on the other platform to keep the forces in balance. To keep such an elevator working indefinitely, there must be readily available people who wish to ride down as many floors as others wish to ride up; their gravitational potential energy is the energy resource required to balance the elevator. With the gravitational forces in balance it will require minimal force to accelerate the elevator in the desired direction. Efficiency that can be achieved by such a mechanism will never reach 100% but it could be very high.

spring-driven elevator

An 'ideal' coiled spring could also be devised to construct an elevator without requiring a motor to drive it as shown in figure 2.3. An ideal spring is one for which the force equation is the following:

$$F = M a = k s$$

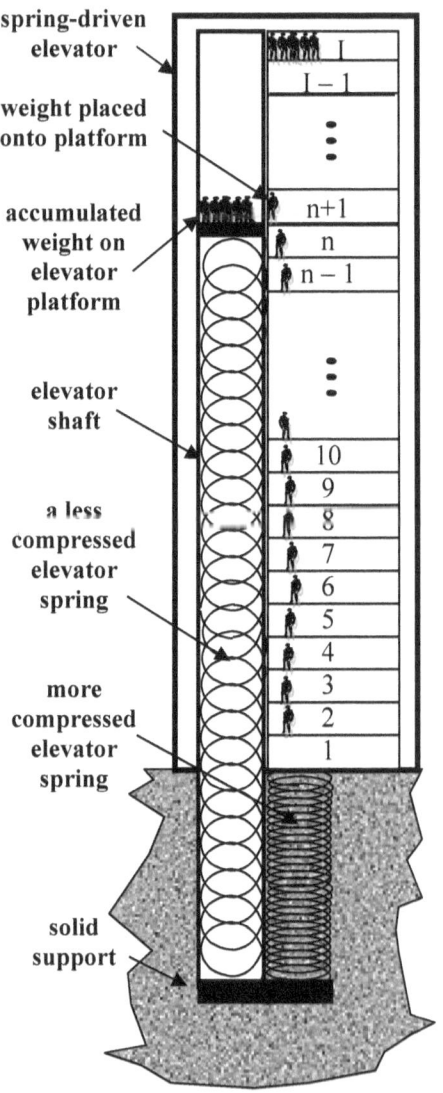

Here again the acceleration, a, is just the force of gravity, g. The mass M is the sum of the masses of incrementally added people who get on the elevator platform. You will notice that we have reduced the size of the people so as to more directly suggest a requirement for 'infinitesimal' units required if we are ever to approximate conditions for which it would be possible for reversibility to apply.

$$M = \sum{}' m$$

The spring constant, k, must be selected to match the separations between floors to the weight of the incremental load of the sum of infinitesimal units m as follows:

$$k = m \, g \, / \, \Delta s,$$

Δs is the floor separation distance.

The procedure is similar to what was discussed with regard to figure 2.2 except that now we have defined the fixed weight-to-floor-separation. Here we encounter the awkwardness of not being able to

Figure 2.3: Analogous spring-driven reversible elevator mechanism

elevate the final incremental weight (person) to the top floor on the return trip and there will be a net accumulation of people who have come down the elevator at the bottom to allow return trips to the top.

a thermodynamic reversible elevator

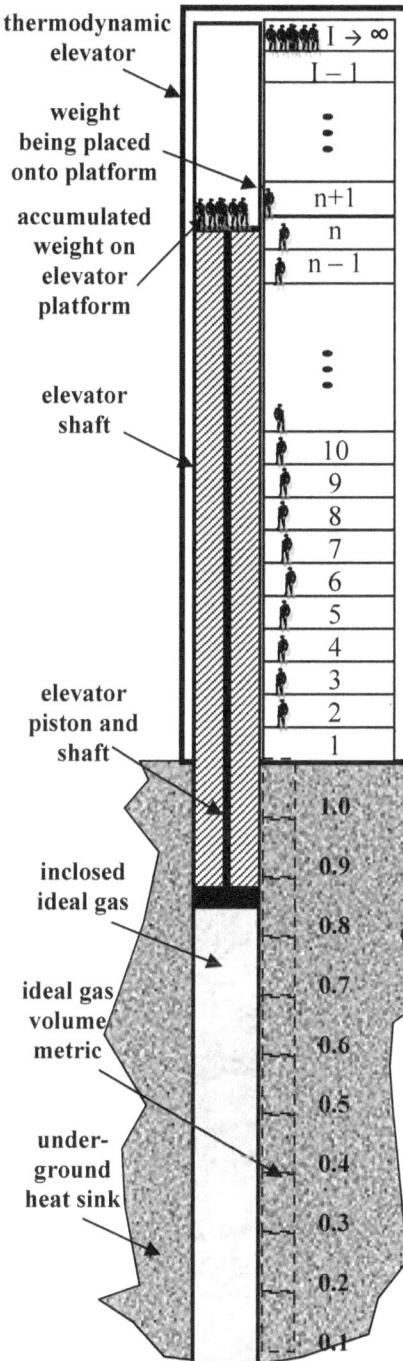

thermodynamic elevator

weight being placed onto platform

accumulated weight on elevator platform

elevator shaft

elevator piston and shaft

inclosed ideal gas

ideal gas volume metric

under- ground heat sink

Figure 2.4: Application of the ideal gas law to a reversible elevator

When a compressed gas expands, forcing a piston to the right against the previous lesser pressure as in figure 1.1, 'work' is performed by the system. Furthermore, idealized 'infinitesimal' changes are reversible to establish more or less perpetuated cycles to be exploited in performing a meaningful cycle of *work* in the everyday sense of that term. If change can be implemented very slowly in infinitesimal steps the work performed will be the ultimate efficiency that is achievable from a thermodynamic system. Admittedly this cannot exploit the total energy invested in such a system, but still, the best that can be realized.

As an example, in figure 2.4 we illustrate an elevator mechanism that exploits a continuous variation of the system state started by work being performed on the system. In this case it is the weight of tiny people used to apply a differential pressure on a piston to compress an enclosed gas to then expand in a 'reversible' cycle as shown in figure 2.5. An elevator shaft along one side of such a structure accommodates a large (infinite in the limit) number of tiny floors. Each floor accommodates a space for tiny people for force-free stepping onto the elevator platform.

Of course, as in the previous mechanical examples such elevators are not without faults. So there are rather severe restrictions on speed and flexibility. Initially the ideal gas shown beneath the piston will be in

19

equilibrium at atmospheric pressure and temperature. The elevator platform is at rest on the top floor to begin operations. A tiny person will step onto the elevator platform adding pressure by his weight to lower the platform to the next floor down where more weight can be added to the platform to proceed further.

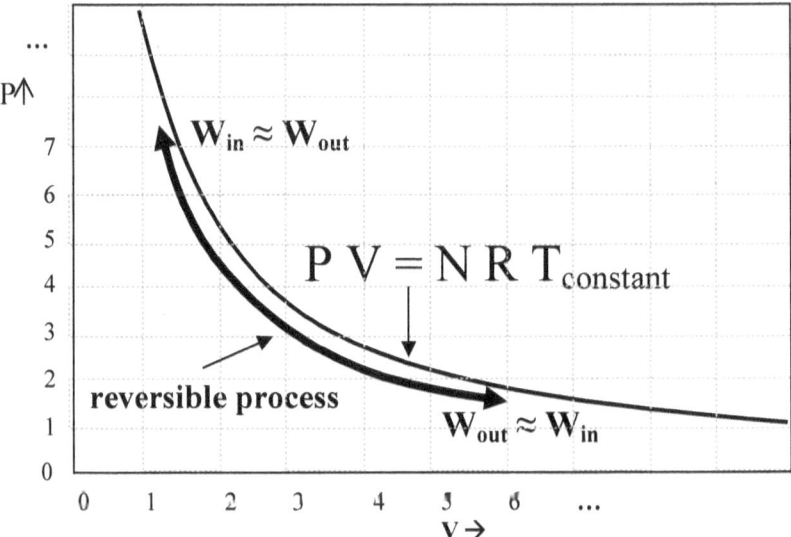

Figure 2.5: Illustration of an isothermal application of the ideal gas law to get work out of a thermodynamic system

Assuming the weights that are added are truly infinitesimal (extremely tiny people), then Newton's laws of mechanics specify that the amount of work performed on or by the elevator can be determined as the integral of the product of the applied force, $F_{down} = m\,g$, times the distance ds over which that force is applied in lowering the elevator one 'infinitesimal' floor at a time; here m is the mass on the elevator platform at a given instant (floor level) and g is acceleration of gravity:

$$W = \int_{d_1}^{d_2} F_{down}\, ds$$

Notice that the pressure is $P_{down} = F_{down}\, A$, where A is the area of the piston. In this case, $F_{up} = P_{up}\, A$, with P_{up} the pressure of the gas and A the area of the piston. Therefore, the incremental distance ds = dV/A, where dV is the incremental change in volume of the gas, so that:

$$W = \int_{V_1}^{V_2} P_{up}\, dV$$

In equilibrium conditions for which the ideal gas law applies, we have the familiar state equation, $P = N R T / V$, where $P = P_{down} = P_{up}$. Thus the amount of work performed will be related to the resulting change in pressure, and therefore, volume of the gas as follows:

$$W = \int_{V_1}^{V_2} N R T \, (1 / V) \, dV$$

$$= N R T \, \ln (V_2 / V_1)$$

This amount of work performed is considerably less than if the resulting expression involved a simple ratio of V_2 / V_1.

design options for thermodynamic elevators

We haven't completely specified the design of our elevator. There are two basically different alternatives for the design of this elevator, either of which could be implemented as reversible processes. To comprehend the alternatives, consider that we are changing the state of the system over time – however slowly – so that taking the time derivative of both sides of the ideal gas law state equation is in order.

$$\frac{d}{dt} (P V) = \frac{d}{dt} (N R T) = N R \, \frac{d}{dt} T = 0$$

It is zero because we are assuming an isothermal process for which the temperature of the gas, T remains constant. Thus we have:

$$P \, \frac{d}{dt} V + V \, \frac{d}{dt} P = 0$$

Hence,

$$P \, \Delta V = - V \, \Delta P$$

The sign difference indicates alternative directions of the increments.

Elevator design options involve a choice of whether a single person placed upon or removed from the elevator platform should move between varied floor separations in the structure or to have varying numbers of people effect constant floor separation as follows:

equal infinitesimal weights alternative

If we choose to have equal weights (i. e., equal ΔP increments) we cannot have the floors equidistant from each other. The upper floors will have much larger separations than the lower floors. This is because, as we saw above, the floor separations are given as $\Delta s = \Delta V / A$ and from the ideal gas law:

$$\Delta V = \frac{V}{P} \, \Delta P = (N R T / P^2) \, \Delta P$$

equal floor height separation alternative

On the other hand, if we choose to have equal spacing between floors, a greater amount of weight will be required at each subsequent lower floor because of the inverse relationship between P and V which produces a nonlinear relationship between increments of pressure (the weight on the platform) and floor levels (proportional to V) as follows,

$$\Delta P = (N R T / V^2) \, \Delta V$$

reversibility using changes in heat rather than pressure

Application of the ideal gas law to a reversible process involving differential changes in temperature rather than pressure as symbolically illustrated in figure 1.2 above is conceptually straight-forward as well. However, in this case it cannot be nearly so directly implemented. Heat would be transferred reversibly between the system itself and insulated surroundings – all taking place very slowly and in infinitesimal steps very like what was required for the reversible mechanical work using pressure. Again there is a differential loss (this time of heat) from the system on each cycle, with a net amount of heat transferring to the surroundings. We will discuss related issues with regard to heat engines in a later chapter.

The analogy with the previous examples is virtually complete. Here we rely on figure 1.3.f rather than on figure 1.3.a to illustrate the analogy as shown in figure 2.6. We choose to use density rather than volume only to stress how directly the analogy applies. Both mechanisms could be used as the workings of a clock similar to a pendulum-driven grandfather clock. Both would require an occasional infusion of additional energy (whether in physical work or in heat) to create the illusion of perpetuity just as a grandfather clock requires a grandfather to periodically wind it to keep it going.

reversibility does not provide perpetuity

In either thermodynamic design option, when the elevator reaches the bottom the weight of the tiny people from all floors 2 through I will be situated on the platform. The elevator is now set up for another run.

To raise the elevator to the next higher level the weight of one tiny person that was added to the elevator from the floor just above where it is now situated must exit onto each successive floor causing the gas to expand, raising the platform to the next floor by reducing the load commensurably. This procedure is repeated until the platform has once again reached the top.

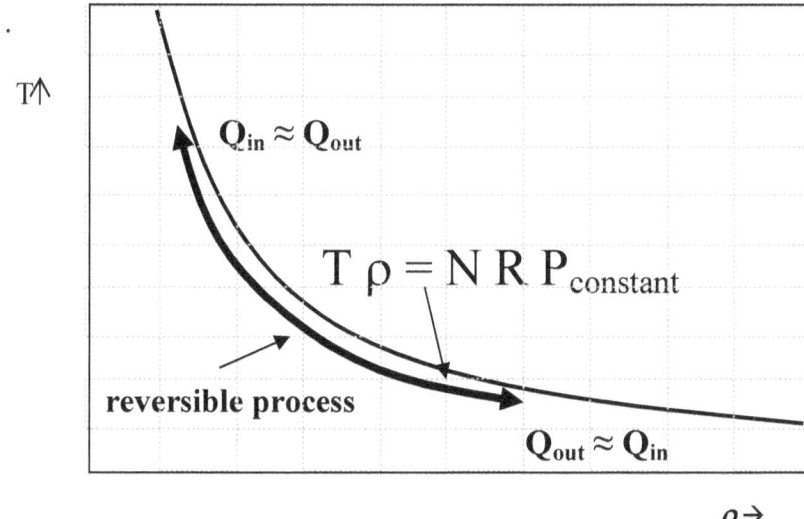

$Q_{in} \approx Q_{out}$

$T \rho = N R P_{constant}$

reversible process

$Q_{out} \approx Q_{in}$

$\rho \rightarrow$

Figure 2.6: Illustration of a constant pressure application of the ideal gas law to get work out of a thermodynamic system

In order for the elevator to begin its descent a second time, another personn will need to be supplied at the top floor to stroll onto the platform, again depleting the number of people available on the top level; there will need to be as many auxiliary people available on the top floor as the number of cycles the elevator will operate. There are alternative procedures for the second and subsequent elevator descents depending on the design we have chosen.

When the elevator with the first design alternative proceeds with its descent, the intermediate floors have had their occupants replenished by the procedures of the previous upward ascent. That is, until floor number 2 is reached. The people from this level had to be left on the bottom floor in order to begin the upward ascent, so there is an accumulation of the people placed on the elevator from higher floors left on the bottom floor. Unless additional weights are supplied on the upper floors, the elevator operation will terminate when it gets back to the top floor.

comments on the nature of reversibility

Basically physics provides models of phenomena that occur in nature and those models allow us to understand the reality around us; it is, in fact, quantitative epistemology. These models accommodate a mathematical representation of natural phenomena that provide an understanding of mundane (or the more obtuse) problems we encounter. But it is first and foremost an epistemological endeavor. However, the approach is limited by the accuracy of the models employed, which in turn is ultimately limited by practical considerations. Later we will discuss the nature of models themselves, the validity of applications of 'open' and 'closed' loop models, with the former requiring a set of assumptions at the boundary of what we define as our 'system' that address those aspects of reality that are included in the model and those that are not. Irreversibility virtually always results from the aspects left out of inaccurate models as invalid assumptions with regard to their boundary conditions.

Classical physics is exclusively about theoretical descriptions of phenomena and processes that can be characterized as reversible. These 'laws' – as they are called – have been codified over the years to treat these simplified cases; they address what are called the 'conservative' forces of nature. Thermodynamics is no different in this regard. If reversibility does not strictly apply in the context of the phenomena of interest then predictions can only be obtained by the use of approximations of the formulations that apply to similar reversible phenomena. Furthermore, as we will illustrate later, the derivations of these formulas that are integral to these 'laws' have ignored second and higher order differential effects so that many of the most challenging problems facing science are not specifically addressed and this includes the 'neglected' effects of motion (and specifically very small differences in relative motion) that accommodate reversible solutions.

There is certainly no wish to disparage the tremendous successes of the scientific method because these successes have indeed been spectacular. We cannot abandon the scientific method or pretend that something else might be better. It isn't! But the more difficult issues – most notably irreversibility – have yet to be adequately addressed by accurate models and it is those requirements that we will address after acknowledging the status quo in the current understanding of thermodynamics.

3: Thermodynamic System Properties

"Thermodynamics does things backward!   We are always generalizing.   But once having the equations we can forget our reversible process (which was but a means to an end) and enter the abstract world of pure mathematics.   You'd never believe what's done with this poor little equation [dU = T dS − P dV] (and all of it exactly right, because it depends only on mathematics).   That's the theoretical half of thermodynamics, and although we will pay some attention to it, we are really more interested in the other half − applied thermodynamics." [1]

In addition to the primary parameters T for temperature, V for volume, and P for pressure that we introduced in conjunction with the ideal gas law, there are quite a few system properties that must be defined. These properties of a thermodynamic system are involved in the mathematical description of various quantitatively defined energies and 'thermodynamic potentials'. They can be variously defined in terms of the primary and other secondary parameters but however defined, they are pertinent to any discussion only to the extent that the associated properties are relevant to processes being considered.

There are two basic types of properties to be considered as follows: An *intensive* physical property of a system is one that does not depend on the size of the system or amount of material in that system.   An *extensive* property on the other hand does depend on the system's size and/or the amount of material in that system.

The following list includes brief definitions of typically encountered properties that will be more fully defined as we proceed.

---

[1] H. C. Van Ness, **Understanding Thermodynamics**, Dover, New York, 1969 (p. 28)
   But, of course we are more interested in the submicroscopic level we save till later.

1. Internal energy   U,
2. Internal heat   Q,
3. Mechanical work   W,
4. Entropy   S,
5. Enthalpy   H,
6. Heat capacity (volumetric) C_V,
7. Heat capacity (constant pressure) C_P,
8. Chemical potential energy μ.

Related parameters of Gibbs, Helmholtz, and Landau 'free energies' in addition to a 'grand potential' are sometimes used in various calculations, but we will forego discussions of them in our brief treatment.

internal energy

The internal energy of a thermodynamic system with well-defined boundary conditions is typically represented by U, although sometimes just E. It is a state function[4] of a system, an extensive quantity. Included is the total of the kinetic energy due to the motion of all the constituent particles contained therein, which involves translational, rotational, and vibrational kinetic energies of the constituents of the system. It also includes the potential energy associated with the vibrational and electric energy of atoms within molecules or crystals of the system. It includes the energy in all the chemical bonds and the energy of free conduction electrons in metals as well. This also incorporates the internal energy of the electromagnetic radiation, notably including blackbody radiation. For a closed system held at constant entropy, this quantity will remain constant.

At the submicroscopic level, internal energy may take on many different forms. It may consist almost entirely of the kinetic energy of gas molecules. It might also consist of the potential energy of these molecules in a gravitational, electric, or magnetic field. For any material, it may also consist of potential energy of attraction or repulsion between the individual constituents of the material.

---

[4] A state function is a quantity that is a property of a system depending only on the current state of that system. It does not depend on the way in which the system got to its current state. It describes an equilibrium state of a system. Internal energy, enthalpy, and entropy are all such state quantities because they quantitatively describe an equilibrium state of a thermodynamic system. Mechanical work and heat, on the other hand, are 'process quantities' because they quantitatively describe the transitions between equilibrium states of thermodynamic systems.

In actuality the internal energy cannot be precisely measured – only changes in this internal energy are measurable as suggested by the first law of thermodynamics discussed earlier where it is seen as dependent on the amount of heat added and work performed by the system:

$$\Delta U = \Delta Q - \Delta P$$

As a matter of definition, the total internal energy of a given system is defined as the difference between the internal energy of that system in its current state and its energy at absolute zero temperature. Since an absolute zero temperature can never be attained, the total internal energy cannot be precisely measured. We will however determine later that for an ideal monatomic gas that, $U = (3/2) N R T$

The typical unit employed for this quantity is the joule or erg depending on the system of units being employed, although the (small or large) calorie of heat is also sometimes employed. The various categories of energy that can go into the composition of *total* internal energy are illustrated in the following table:

Type	Composition of Internal Energy (U)
Sensible energy	This includes that portion of the internal energy of a system associated with kinetic energies: translational, rotational, vibrational, and spin energies associated with the molecules, nuclei, and electron constituents.
Latent energy	This is the internal energy associated with the phase of a system.
Chemical energy	This corresponds to the internal energy associated with the chemical bonds in the molecules.
Nuclear energy	This is the huge energy associated with the strong nuclear force that binds the nuclei of the atoms.
Interaction energy	This includes those types of energies not stored in the system directly such as heat transfer, mass transfer, and mechanical work. These energies cross system boundaries as gains or losses by the system during a physical process.
Thermal energy	This is the sum of sensible and latent forms of internal energy.

internal Heat

The symbol Q is used for the amount of heat added to or removed from a system (measured in joules or ergs). A positive value for Q represents heat flowing into the system while a negative value indicates that the heat is flowing out of the system. What this internal 'heat' *is* remains somewhat obscure – whether totally accounted for by the motions of submicroscopic particulate matter or more obscurely by associated thermal radiation or both.

mechanical Work

The symbol W represents the mechanical work done on a system (also measured in joules or ergs). It is the energy added or extracted by the various mechanical contrivances we will discuss in the next chapter. Needless to say, it requires physical 'work' to compress a gas, and work can be done by allowing a gas to expand.

entropy

Entropy S is an extensive parameter of a system that helps to assess the flow of energy through a thermodynamic process. It quantifies the amount of a system's thermal energy that is unavailable for conversion into mechanical work. Entropy can be defined at both the macroscopic and submicroscopic levels, but rather obscurely in each.

It was originally defined at the macroscopic level for what were called thermodynamically reversible processes, i. e., those for which, after each cycle, the system is returned to virtually the same state as we have seen. Thus, by reversing the equation for an incremental heat change, we obtain:

$$dS = dQ / T$$

where the temperature T is divided into the increment of heat energy dQ that is transferred into and out of a system on each cycle in performing mechanical work. As is usual with macroscopic thermodynamics, this definition ignores whatever the associated facts might be concerning the lower levels of reality involving the atoms and molecules of which the system is comprised.

In thermodynamics, entropy has been found to be very generally useful and it has several other reformulations. Entropy was discovered via mathematics, which merely states that it was derived by analogy to the role of volume in the determination of mechanical work. It is understood more as a logical consequence of relations between physical parameters that were discovered by experiment than as having itself resulted from some particular experiment. It is a quantity that behaves as a function of the overall state of a system.

At a lower level of description in statistical mechanics, the additional concepts of 'order' and 'disorder' of lower level constituents of a system were introduced into the concept of entropy as additional concomitants as we will discuss in more detail further on. Recently thermodynamic processes are often described in terms of the state of order of a system, such that entropy has become an expression of the degree of randomness. In more modern submicroscopic interpretations entropy is defined as the amount of additional information needed to specify the exact physical state of a system, given the system's higher level thermodynamic description in terms of the usual pressure, volume, temperature, and density. As a consequence, the second law of thermodynamics is now more typically seen by physicists as dependent on this particular definition of entropy.

In this scheme entropy is more descriptively expressed in the following terms:

$$S \equiv k + \log n$$

Where k is Boltzmann's constant as previously defined and n is defined as the number of submicroscopic particle states in the system consistent with a single given state of the system at the macroscopic level. There is, of course, a huge number of alternative possible positions and velocities of the individual particles in the phase space of the system. All of these possible alternatives give rise to the very same volume, pressure, density, and temperature (macroscopic state) of that system but with an effectual difference with regards entropy.

We will expend considerable discussion on this topic presently.

changes in internal energy

The first law of thermodynamics introduces the concept of the conservation of energy, both mechanical and thermal. Thus changes in the internal energy of a closed system must be equal to the sum of the amount of heat energy Q supplied to the system minus the amount of mechanical work W done by the system.

The internal energy is defined by the first law of thermodynamics which states that energy is conserved:

$$\Delta U = Q - W,$$

ΔU is the net change in internal energy of a system during infinitesimal changes of a process, so the first law may be equivalently expressed as:

$$dU = \delta Q - \delta W,$$

where the terms now represent infinitesimal amounts of the respective quantities. The 'δ' before the internal energy function indicates an exact differential. Thus, U is a state function associated with the system. The

Greek δ's before terms, on the other hand, indicate that rather than describing increments of energy that are state functions, they pertain to processes by which internal energy is changed.

But frankly, this is more typically just expressed as:

$$dU = dQ - dW$$

These two components of the internal energy may then be expressed in terms of other thermodynamic parameters. Each term in the associated expression will be composed of an 'intensive variable' (which will be a generalized force) and a conjugate 'infinitesimal differential' of another 'extensive variable' that acts as a generalized displacement.

For a non-viscous fluid, mechanical work done by or on a system will be reflected in the associated pressure P and volume V of the system. In the following expression for work performed, analogous to a force being applied over a displacement distance, the pressure acts as the intensive generalized force, $P = F \ A$. The change in volume acts as the extensive generalized displacement, $ds = dV/A$, so that:

$$dW = P \ dV$$

This assumes the default direction of mechanical work, W, done by the system as that which proceeds from the working substance to its surroundings. Similarly, if we take the default direction of transfer of *internal heat*, Q, to be into the substance doing the work, we obtain:

$$dQ = T \ dS$$

This assumes a 'reversible' process for which 'entropy S is conserved.

Using the above expressions and plugging them back into the first law of thermodynamics we can construct a possible expression for the internal energy of a closed system, whether directly measurable or not, as follows:

$$dU = dQ - dW = T \ dS - P \ dV$$

The internal energy function may be written as U(S,V), i. e., as a function of the entropy and volume of the system. And since U, S, and V are all extensive parameters:

$$U(\alpha S, \alpha V) = \alpha \ U(S,V),$$

where α is a scalar constant. This implies that $U(S,V)$ is a homogeneous function throughout the system. Thus, from Euler's homogeneous function theorem we obtain the typical expression for internal energy as:

$$U = T\,S - P\,V$$

For an elastic substance the pressure must be replaced by more complicated tensor expressions for the associated stresses and strains within the system, but these are complications we will not explore.

If a (non-viscous) fluid gains energy by the addition of more particles, we add the chemical energy term for the additional particles:

$$dU = T\,dS - P\,dV + \mu\,dN$$

and again,

$$U = T\,S - P\,V + \mu\,N$$

where μ is the chemical potential energy of the chemical species that are added. It is an 'intensive variable'. N is the number of particles of the chemical species that has been added. N is an extensive variable. Note that a separate term of the form $\mu_i\,N_i$ must be added for each unique chemical species i that is incorporated into the system.

enthalpy

Enthalpy is the internal energy of the system plus the energy related to pressure-volume work done by the system.

$$H \equiv U + P\,V = U + N\,R\,T$$

heat capacity

There are two quantities that are metrics for the ability of a substance to retain heat. For a given volume of a gas, as a matter of definition,

$$C_V \equiv \left(\frac{\partial U}{\partial T}\right)_V ,$$

which is the rate of change of the internal energy of a gas with respect to changes in temperature when the volume of the gas is held constant.

C_V is called the heat capacity of the gas at constant volume or *volumetric heat capacity*. It is always nonzero.

We also have by definition that for a given pressure:

$$C_P \equiv \left(\frac{\partial H}{\partial T}\right)_P$$

It is defined in a similar sense to the definition of C_V.

For an ideal monatomic gas with three degrees of freedom the internal energy is:

$$U = (3/2)\, N\, R\, T$$

The change in internal energy as a function of temperature with a constant volume is:

$$\Delta U = (3/2)\, N\, R\, \Delta T = C_V\, \Delta T$$

And since $H = U + N R T$, with constant pressure we obtain:

$$\Delta(P\ V) = N\ R\ \Delta T,$$

So that:

$$\Delta H \equiv C_V\, \Delta T + N\, R\, \Delta T,$$

Thus, we arrive at $C_V = (3/2)\, N\, R$ and $C_P = (5/2)\, N\, R$. Their ratio γ is:

$$\gamma \equiv \frac{C_P}{C_V} = \frac{5}{3}$$

This applies to an ideal monatomic gas to which the ideal gas law restrictions pertain. In particular, the value of this ratio holds very precisely for the noble gases, i. e., helium, neon, argon, etc..

4: Thermodynamic Processes

"The idea underlying this concept is that in a gas, heat is nothing else than the kinetic or mechanical energy of motion of the gas molecules, so that, in expanding, a gas does work at the expense of kinetic energy of its molecules, which represent its heat energy." [1]

We investigated implications of changes to the equilibrium status of an ideal gas in the last chapter. Now let's consider some experiments with real gases. For example, Joule designed an experiment from which to determine whether, and to what extent, gases cool as they expand. He insulated an apparatus of two sealed glass bulbs that were connected but had a stopper that could be opened or shut between them. The experiment involved filling one bulb with a gas at a particular pressure P and temperature T, leaving the other bulb evacuated, i. e., P near zero. The insulation precluded heat from flowing into or out of his apparatus such that $\Delta Q = 0$, i.e., no heat could enter or exit.

isothermal expansion

When the stopper was opened between the containers, it allowed gas to expand into both containers, thus increasing the volume of the gas and reducing its pressure. With no pressure to resist its flow into the previously evacuated container, there was no work performed by the expansion so that $\Delta W = 0$. Since both ΔQ and ΔW were zero, there was no net change in the internal energy of the gas:

$$\Delta U = \Delta Q + \Delta W = 0.$$

---

[1] Leonard Loeb, *Kinetic Theory of Gases*, McGraw-Hill, New York, 1927, p. 13

After the volume had increased by the gas expanding into the second container, Joule demonstrated that there had been no change in the temperature. In this way he determined that at least to the accuracy of his measurement for the gas he had used (and in general for an ideal gas), there is no change in the temperature in response to such a change in volume of a gas. So that from the ideal gas law where we have that:

$$P = (N \, R \, T) \, / \, V,$$

the entire parenthetical expression is constant in this case. Implied isothermal expansion-compression curves were presented in figure 1.3.a and are illustrated as the solid curves plotted on the pressure versus volume graph in figure 4.1. This is similar to the prediction for the reversible process example illustrated in figure 2.5, since overall stasis is assumed in the applicability of the ideal gas law in both cases.

Thus, what Joule had discovered based on his experiment was that:

$$\left(\frac{\partial T}{\partial V} \right)_U = 0$$

Here the expression on the left is the rate of change of temperature T with respect to the rate of change in the volume V of a gas, with the constraint that the internal energy U be held constant. Using Euler's chain relation of derivatives, we have:

$$\left(\frac{\partial T}{\partial V} \right)_U = \frac{\left(\dfrac{\partial U}{\partial V} \right)_T}{\left(\dfrac{\partial U}{\partial T} \right)_V}$$

But as a matter of definition of heat capacity we saw that,

$$C_V \equiv \left(\frac{\partial U}{\partial T} \right)_V$$

where C_V is the heat capacity of the gas at constant volume, which is always finite and nonzero. So as a consequence we have also that:

$$\left(\frac{\partial U}{\partial V} \right)_T = 0$$

Taken together, these equations imply that the internal energy U of an ideal gas is exclusively a function of temperature T and thus we arrive at the more general statement concerning the internal energy of an ideal gas, namely that:

$$U = U(T)$$

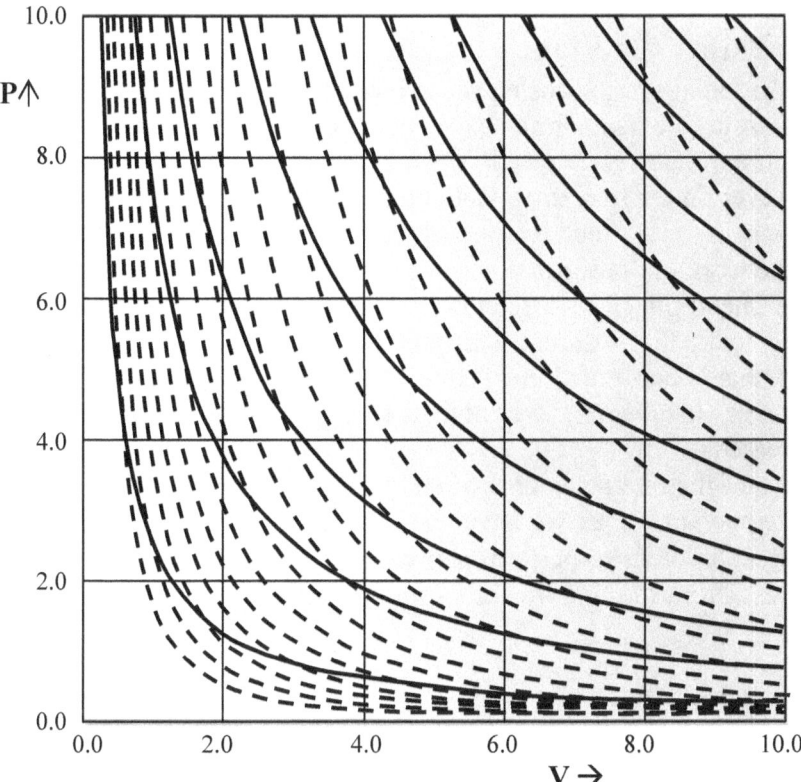

Figure 4.1: Isothermal and adiabatic expansion-compression curves

More accurate measurements with real gases, reveal that this is not precisely the case for all gases. Nonetheless, this rate of change of internal energy with volume is always very small.

This exclusive dependence of internal energy on temperature can be extended further to include the enthalpy H of ideal gases as well. And since,

$$H = U + P\,V$$

And because $P\,V = N\,R\,T$ for an ideal gas, we must have that:

$$H = H(T)$$

Thus, for ideal gases both U and H are exclusively functions of T and all of the following partial derivatives vanish:

$$\left(\frac{\partial U}{\partial V}\right)_T = \left(\frac{\partial U}{\partial P}\right)_T = \left(\frac{\partial H}{\partial V}\right)_T = \left(\frac{\partial H}{\partial P}\right)_T = 0$$

adiabatic expansion

An 'adiabatic expansion' is defined as one that requires that $\Delta Q = 0$ similar to the isothermal expansion constraint discussed above for which no heat is allowed to enter or exit the system under test. However, in contrast to an isothermal expansion, changes of state from work being performed by the system are allowed; the gas is allowed to do the work on its surroundings such that $\Delta W \neq 0$. Since, as in Joule's experiment, the gas is insulated from any external source of heat, it can not convert that external heat into mechanical work. Thus, any such work that is performed must come at the expense of the internal energy of the gas itself so that the internal energy U of the gas will thereby be decreased.

Since the internal energy of an ideal gas is exclusively dependent on the temperature T as we have shown, the temperature of the gas must decrease in order to perform any work. Since $\Delta Q = 0$ for an adiabatic process, from the first law of thermodynamics, we must therefore have that:

$$\Delta U = - P \, \Delta V$$

Thus, we have in this case that U is a function of both T and V, since it changes with changes in V as well as T, so that for an adiabatic process we must have that:

$$U = U(T,V)$$

And furthermore, since,

$$\Delta U = \left(\frac{\partial U}{\partial T}\right)_V \Delta T + \left(\frac{\partial U}{\partial V}\right)_T \Delta V,$$

Since $U = (3/2) \, N \, R \, T$ for an ideal monatomic gas, it follows that,

$$\Delta U = C_V \, \Delta T$$

by the definitions of C_V and implication of an ideal gas for which the second term vanishes because U cannot change without a change in T .

Expressions in the right hand sides of the two equations for ΔU must be equal so that:

$$- P\, \Delta V = C_V\, \Delta T$$

And therefore as applicable to the ideal gas law,

$$C_V\, \Delta T = - N\, R\, T\, \Delta V\, /\, V$$

By rearranging factors and replacing the 'difference' symbol Δ with the similar infinitesimal differential, d, appropriate to the integral calculus, we obtain:

$$\frac{dT}{T} = -\frac{N\,R}{C_V}\frac{dV}{V}$$

Applying a similar approach to enthalpy, we see that since $\Delta Q = 0$,

$$\Delta H = V\, \Delta P$$

And again, since

$$\Delta H = \left(\frac{\partial H}{\partial T}\right)_P \Delta T + \left(\frac{\partial H}{\partial P}\right)_T \Delta V,$$

We have that,

$$\Delta H = C_P\, \Delta T$$

Here C_p is heat capacity measured at constant pressure.

Expressions in the right hand sides of the two equations for ΔH must be equal so that:

$$V\, \Delta P = C_P\, \Delta T$$

And again, as applicable to the ideal gas law,

$$C_P\, \Delta T = N\, R\, T\, \Delta P\, /\, P$$

Rearranging factors and replacing the symbol Δ with the infinitesimal differential, d, appropriate to the integral calculus, we obtain:

$$\frac{dT}{T} = \frac{N\,R}{C_P}\frac{dP}{P}$$

Equating the two expressions for dT/T, we see that:

$$-\frac{N\,R}{C_V}\frac{dV}{V} = \frac{N\,R}{C_P}\frac{dP}{P}$$

Simplifying, we obtain:

$$-\frac{C_P}{C_V}\frac{dV}{V} = \frac{dP}{P}$$

So that:

$$\frac{dP}{P} = -\gamma\,\frac{dV}{V}$$

where we have used the earlier definition:

$$\gamma \equiv C_P / C_V$$

As long as the two specific heats of the gas remain constant we can integrate the previous difference equation to obtain:

$$\ln P = -\gamma \ln V + \ln K_1$$

Exponentiation of both sides obtains:

$$P = K_1\, V^{-\gamma}$$

Here K_1 is a constant of integration. So when integrated between system states denoted 1 and 2 we obtain:

$$\ln\frac{P_2}{P_1} = \gamma \ln \frac{V_1}{V_2}$$

Thus, we arrive at the relation:

$$P_1\, V_1{}^{\gamma} = P_2\, V_2{}^{\gamma}$$

These equations apply to the adiabatic expansion-compression of an ideal gas. By inspection of figure 4.1, where adiabatic curves are drawn as dashed lines on the pressure versus volume graph, one can readily determine that for $\gamma \neq 1$, these adiabatic curves (shown as dashed lines) must necessarily cross the isothermal lines, so indeed temperature is affected by such expansion and compression although the internal energy of the system remains unchanged.

It is instructive to consider the plot of isothermal and adiabatic expansion/compression curves on a log scale for both P and V as shown in figure 4.2

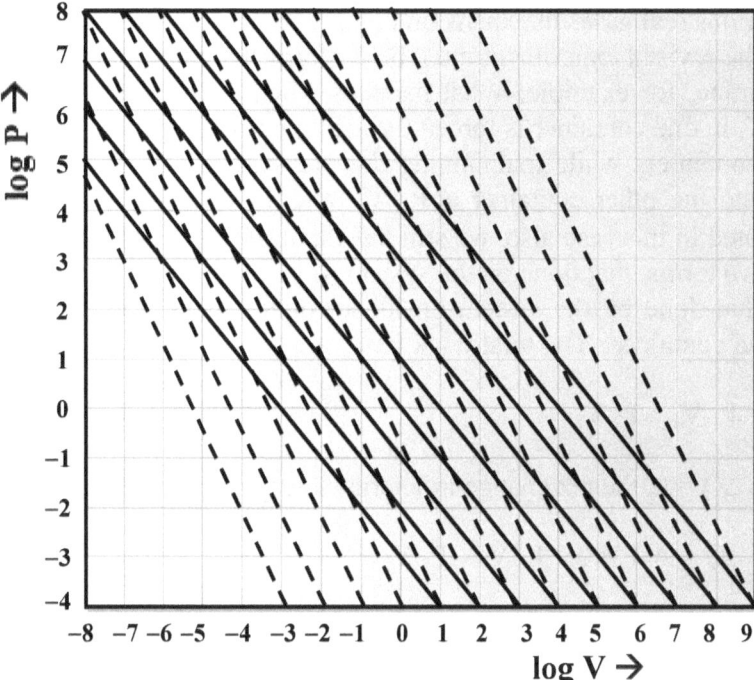

Figure 4.2: Isothermal and adiabatic expansion/compression curves (log scale)

With the expression above for P(V), the work performed by the system during adiabatic expansion can be calculated as follows:

$$\Delta W = - P(V) \, \Delta V$$

So that,

$$W = -\int_{V_1}^{V_2} P(V) \, dV$$

Resulting in the expression:

$$W = - \frac{K_2}{\gamma + 1} \left(\frac{1}{V_2^{\gamma-1}} - \frac{1}{V_1^{\gamma-1}} \right)$$

The constant K_2 is fully determined by a single point on the adiabatic path followed by the expansion-contraction.

Joule-Thompson expansion

The results of Joule's isothermal expansion experiment are not completely valid for real gases. More accurate experiments were carried out by Joule and J. J. Thompson to more fully understand the properties of various real gases in expansion-compression experiments.

The revised experiment must be performed somewhat differently to determine, for example: What happens when a gas, initially at P_1, V_1, and T_1 in one container is forced through a porous partition between the two containers while maintaining the pressure at P_1? The gas comes out into the other container at P_2, V_2, and T_2. Since the apparatus is insulated in this case also, we still have that $\Delta Q = 0$. But the work now has two terms, that done on the system to force the gas through the plug and that done by the system on its surroundings as it comes into the second container. The total work is:

$$\Delta W = P_1 V_1 - P_2 V_2$$

Since $\Delta Q = 0$, the net change in internal energy of the gas is,

$$\Delta U = \Delta Q + \Delta W = 0 + P_1 V_1 - P_2 V_2$$

So, unlike the Joule expansion, this process does not involve a constant internal energy. However, since the change in enthalpy is given by,

$$\Delta H = \Delta U + \Delta(P\ V)$$

ΔH will be zero in this case. The Joule-Thompson experiment involves, therefore, a process performed at constant enthalpy. The experimental apparatus is such that a change in pressure ΔP can be imposed followed by a measurement of ΔT. The ratio of these two quantities approximates the derivative,

$$\frac{\Delta T}{\Delta P} \cong \left(\frac{\partial T}{\partial P}\right)_H = \mu_{JT}$$

μ_{JT} is a 'coefficient of the Joule-Thompson effect'. Its value would always be zero for an ideal gas but nonzero for any real gas. If we apply the Euler chain rule again to the previous equation we obtain the following:

$$\mu_{JT} = \left(\frac{\partial T}{\partial P}\right)_H = - \frac{\left(\frac{\partial H}{\partial P}\right)_T}{\left(\frac{\partial H}{\partial T}\right)_P} = - \frac{\left(\frac{\partial H}{\partial P}\right)_T}{C_P}$$

Although the numerator in this equation vanishes for an ideal gas, it does not necessarily do so for real gases.

The coefficient of the Joule-Thompson effect is important in the determination of the liquefaction behavior of gases; it is what determines whether a gas will cool or heat upon expansion. It is a decreasing function of temperature, passing through zero at 'the Joule-Thompson inversion temperature', T_I.   During expansion ΔP will always be less than zero, but whether ΔT is positive or negative depends on the sign of μ_{JT}. Looking at the definition of μ_{JT}, we see that if it is positive then ΔT will be negative upon expansion such that the gas will cool. However, if μ_{JT} is negative, ΔT will be positive with the gas warming upon expansion.

In order to liquefy a gas by Joule-Thompson expansion the gas must first be cooled below its inversion temperature. These inversion temperatures vary over a wide range for typical gases. For example: The inversion temperature for O_2 is 764 K, for N_2 it is 621 K, but for He it is the extremely cool 40 K, and for Ne a slightly warmer 231 K.

compressibility of real gases

'Compressibility' is a term used in thermodynamics to describe the differences between the properties of *real* gases and those that would be realized for an *ideal* gas. The compressibility of a substance is a measure of the relative change in volume in direct response to a change in pressure applied to it. The *compressibility factor F* is defined as:

$$F \equiv \frac{P\,V}{N\,R\,T}$$

For an ideal gas this compressibility factor would be equal to unity, and the equation would merely be the familiar ideal gas law. However, a more general quantitative compressibility parameter K is defined as:

$$K \equiv -\frac{1}{V}\frac{\partial V}{\partial P}$$

isothermal compressibility

A possible constraint for performing this measurement is to maintain a constant temperature during the change in pressure. The response then reflects what is referred to as isothermal compressibility. Since by the ideal gas law $P \cdot V$ would be constant at constant temperature, we should have:

$$\left(\frac{\partial(P \cdot V)}{\partial P}\right)_T = 0$$

So that we should therefore have:

$$V = - P \left(\frac{\partial V}{\partial P}\right)_T \text{, and therefore,}$$

$$K_T = \frac{1}{P} \text{, for the isothermal case.}$$

K_T determines how a 'real' as against an 'ideal' gas responds to changes in pressure if the overall temperature remains constant.

The speed of sound in a gas depends on how a volume of the gas responds to changes in its pressure. Newton assumed that the speed of sound involved such an isothermal process and used K_T to calculate the speed of sound in gas. His calculation did not agree with experiments. Thus, the transmission of sound in a gas is *not* an isothermal process; it must be adiabatic instead. The oscillations in pressure produced by sound waves are sufficiently rapid that heat cannot flow from compressed regions before the gas in that region has become rarefied again. And before heat is conducted away from that region the sound compression has propagated to the next depression on its path.

adiabatic compressibility

Adiabatic processes involve no net change in internal heat Q as well as constant entropy, so we define adiabatic compressibility as follows:

$$K_S = - \frac{1}{V} \left(\frac{\partial V}{\partial P}\right)_S.$$

This can be evaluated in terms of known quantities using Euler's cyclic rule twice, once normally and once in reverse:

$$K_S = + \frac{1}{V} \frac{\left(\frac{\partial S}{\partial P}\right)_V}{\left(\frac{\partial S}{\partial V}\right)_P} = + \frac{1}{V} \frac{\left(\frac{\partial S}{\partial T}\right)_V \left(\frac{\partial T}{\partial P}\right)_V}{\left(\frac{\partial S}{\partial T}\right)_P \left(\frac{\partial T}{\partial V}\right)_P}$$

Then, since

$$C_V = T \left(\frac{\partial S}{\partial T}\right)_V$$

and

$$C_P = T \left(\frac{\partial S}{\partial T}\right)_P$$

We have that:

$$K_S = \frac{C_V}{C_P} \frac{1}{V} \frac{\left(\frac{\partial T}{\partial P}\right)_V}{\left(\frac{\partial T}{\partial V}\right)_P} = -\frac{C_V}{C_P} \frac{1}{V} \left(\frac{\partial V}{\partial P}\right)_T$$

And thus we come to the relation:

$$K_S = \frac{C_V}{C_P} K_T$$

Since $C_p > C_V$, we see that the isothermal compressibility will always be greater than the adiabatic compressibility. For an ideal monatomic gas, $C_p = (5/2) N R$ and $C_V = (3/2) N R$ so that in that case,

$$K_S = \frac{3}{5} K_T$$

This was evident in figures 4.1 and 4.2 where both isothermal and adiabatic compression-expansion curves are drawn.

5: The motive power of heat

"Thus all heat engines convert only a part of their heat intake into work and discard the remainder to the surroundings.   This limitation on heat engines is not contained within the First Law of Thermodynamics.   Nor does it result from imperfections in the engines, for we have found it by examining processes carried out reversibly, that is, as perfectly as we can imagine.   This suggests that there must be a Second Law of Thermodynamics which imposes limits not expressed by the First Law."   – H. C. Van Ness (.p. 40)

In the development of steam engines and later combustion engines, it is in the expansion and compression of gases in cycles involving the heating and cooling of those trapped gases that provides the conversion of heat to useful mechanical work.   Although steam engines were invented much earlier than the internal combustion engine, we will begin with a discussion of the basic operation of combustion engines.

internal combustion engines

An internal combustion engine involves a cylinder of a gas (air) with a piston at one end connected via a rod to a crank shaft and flywheel that provides rotary motion when the piston moves back and forth.   That is the basic mechanical design as illustrated in figure 5.1. But it is what transpires within the cylinder that is the ultimate source of the work that such engines perform.   That is where practical application of thermodynamics affects us all most dramatically.

For discussion, let's assume that the piston slides up and down in the cylinder without friction with the volume of the enclosed gas (air mixed with gasoline vapor through the first half of the cycle and carbon oxides and steam in the last half) expanding and compressing with the pressure applied or released by the piston. This also assumes perfectly tight seals around and above the piston.

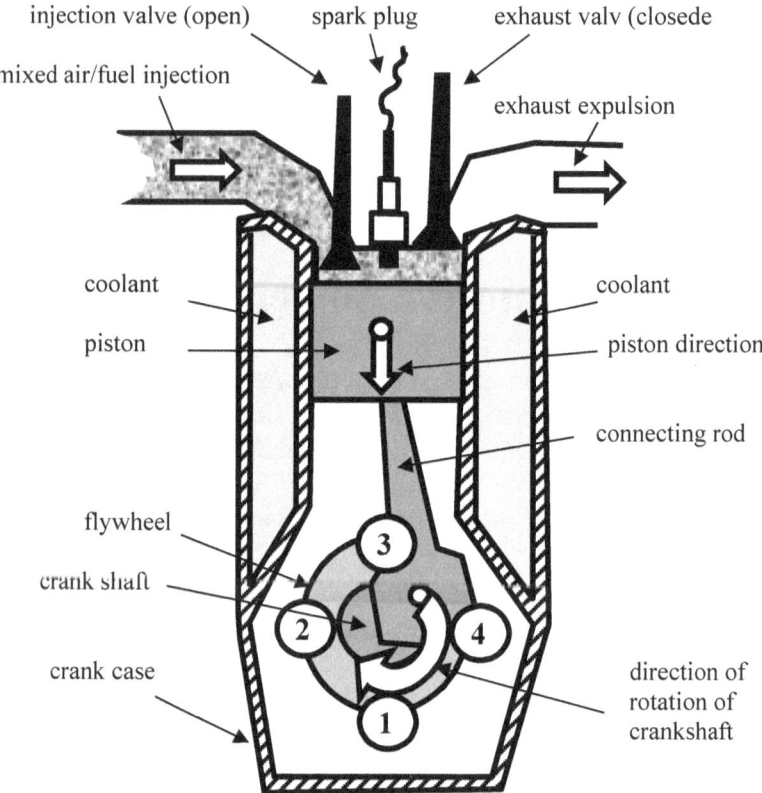

Figure 5.1: Mechanical aspects of an internal combustion engine

When an engine is running, gases in the cylinder undergo a series of changes in repeatable steps in a continuous cycle as shown in figure 5.2. We characterize the process and state of the thermodynamic system at each of steps 1 through 4. Step 0 is auxiliary to the thermodynamics.

Step 0 → 1: Injection

The injection valve is opened to allow a mixture of air and gasoline vapor to be injected into the cylinder at approximately atmospheric pressure. Then as the piston moves to expand the volume in the cylinder from its minimum, it is filled with the fuel mix. When the volume has reached its maximum, the injection valve will be closed. Step 1 is assumed to be at one atmosphere pressure:

$P_1 = 1$ and $T_1 = 600$ K.

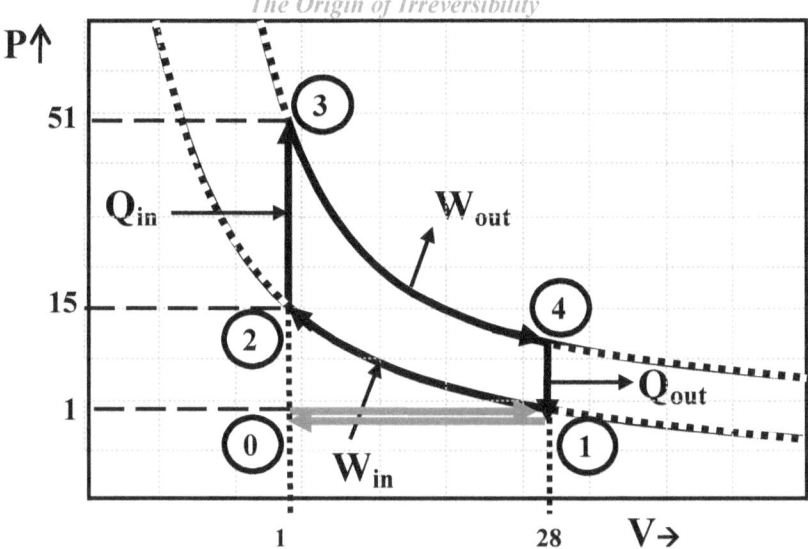

Figure 5.2: Steps in the thermodynamics of a combustion engine cycle

Step 1 → 2: Compression

As a part of the continual rotary motion of the flywheel, inertia will force the piston to compress the trapped air/gasoline mixture in the cylinder. This compression is an adiabatic process with no heat added. We assume a compression ratio of 15 from which we can calculate the temperature using formulas derived earlier as follows:

$$P_2 = 15 \text{ and } T_2 = T_1 \left[\frac{P_2}{P_1} \right]^{(\gamma-1)/\gamma} = (600)(15)^{0.286} = 1300 \text{ K, since}$$

$\gamma = C_P / C_V = 7/5 = 1.4$ for ideal gases.

Step 2 → 3: Combustion

When the gases in the cylinder are near the maximum compression, heat is added very rapidly by igniting the mixture with a spark from the spark plug. The added heat introduced by the sudden combustion of the mixed gasoline vapor and air will increase the temperature dramatically and with it the pressure:

$$T_3 = 4400 \text{ K and } P_3 = P_2 \left[\frac{T_3}{T_2} \right] = (15)(3.85) = 51$$

Step 3 → 4: Expansion

The carbon oxides and water vapor products of the combustion expand forcing the piston back out of the cylinder and because:

$V_3 = V_2$ and $V_4 = V_1$,

we have:

$$T_4 = T_3 \left[\frac{V_3}{V_4} \right]^{(\gamma-1)} = T_3 \left[\frac{V_2}{V_1} \right]^{(\gamma-1)} = T_3 \left[\frac{T_1}{T_2} \right] = (4400)\,(0.46) = 2030 \text{ K}$$

Step 4 → 0: Expulsion

When the expansion has reached its maximum extent, the exhaust valve will be opened dropping the pressure back to one atmosphere. The piston will continue on its compression cycle, but with the exhaust valve open there will be no change in pressure. This allows the combustion products to be completely expelled after which the exhaust valve will be closed. The piston and cylinder are cooled as rapidly as possible by water and or cool air external to the cylinder. The engine is ready for its next cycle.

combustion engine efficiency

Although the gases in the chamber of the cylinder are changed at each cycle using the exhaust and injection valves between steps 1 and 0 and 0 and 1, respectively, we assume a mere change in the state of the same ideal gas. And indeed, from one cycle to the next this characterizes the situation fairly well. We have, therefore also assumed 'reversibility' of the processes between steps in the same sense to which we have become accustomed. This latter assumption is valid as the upper limit of the efficiency of such an engine. To make an assessment of this upper limit of efficiency, we have adapted a discussion by Van Ness using the equations we developed in the previous chapters. [5]

From our previous derivations, we see that the temperature values at each step are all that is needed to calculate the amount of heat required and produced as well as the amount of mechanical work required and produced by an engine. Continuing with our assumption of an ideal gas, C_V will have a constant value throughout. As discussed earlier, the change in the internal energy of a system is just:

$$\Delta U = Q - W$$

And we have also

$$\Delta U = C_V \, \Delta T$$

---

[5] H. C. Van Ness, (p. 32-33)

so that:

$$C_V \, \Delta T = Q - W$$

In the step-by-step discussion above, we see that either Q or W is exactly zero in the processes between steps. In the following calculations we use British thermal units (BTU) for which $C_V = 5$.

$$W_{in} = -5 \,(1300 - 600\,) = -3,500$$

$$Q_{in} = 5 \,(4400 - 1300\,) = 15,500$$

$$W_{out} = -5 \,(2030 - 4400\,) = 11,850$$

$$Q_{out} = -5 \,(600 - 2030\,) = -7,150$$

The total work produced in one cycle is:

$$W_{cycle} = 11,850 - 3,500 = 8,350$$

At each cycle 15,500 Btu of heat must be added to obtain this amount of work, so the efficiency η of the engine is:

$$\eta = \left(\frac{W_{cycle}}{Q_{in}} \right) = 0.54$$

Only 54% of the energy added as heat is converted to mechanical work. So even with the assumption of an ideal gas with reversible processes, 46% of the energy is lost. In 'real' as against idealized engines much more is lost. Van Ness estimates that the best efficiency obtainable on compression and expansion processes is about 75%. This means that instead of 3,500 Btu required between steps 1 and 2, a total of 4,670 would be required. Heat will be dissipated so that $Q_{12} \neq 0$ as assumed:

$$Q_{12} = \left(\frac{W'_{in}}{0.75} \right) + C_V \, \Delta T_{12} = -4,670 + 5 \,(1300 - 600) = -1170$$

Since that much heat will be transferred to the cylinder during the compression process, it cannot be considered adiabatic.

Similarly to the expansion process from which the net positive mechanical work is produced we must apply the 75% reasonableness restriction from which we obtain:

$$Q_{34} = 0.75 \, W'_{out} + C_V \, \Delta T_{34} = 8,900 + 5 \,(2030 - 4400) = -2,950$$

The amount of heat transferred to the cylinder is higher than for the compression process because the internal gases are hotter. So the total work required and produced by the 'real' engine is:

$$W'_{cycle} = W'_{out} - W'_{out} = 8{,}900 - 4{,}670 = 4{,}230$$

So the efficiency of an actual engine will be more on the order of:

$$\eta = \left(\frac{W'_{cycle}}{Q_{in}} \right) = 0.27$$

heat engines more generally

Thermodynamics as a quantitative science developed in large part from the technological need to increase the efficiency of early steam engines, which are, of course, *external* combustion engines. Theory concerning exchanges of energy between quite mundane mechanical 'work' and more ethereal 'heat' were of particular interest in this regard. In 1697, Thomas Savery built the first practical steam engine. Because of obvious commercial application, the study of such devices attracted funds and, therefore, the attention of many of the leading scientists of the time. Figure 5.3 illustrates a simplified design of a steam engine.

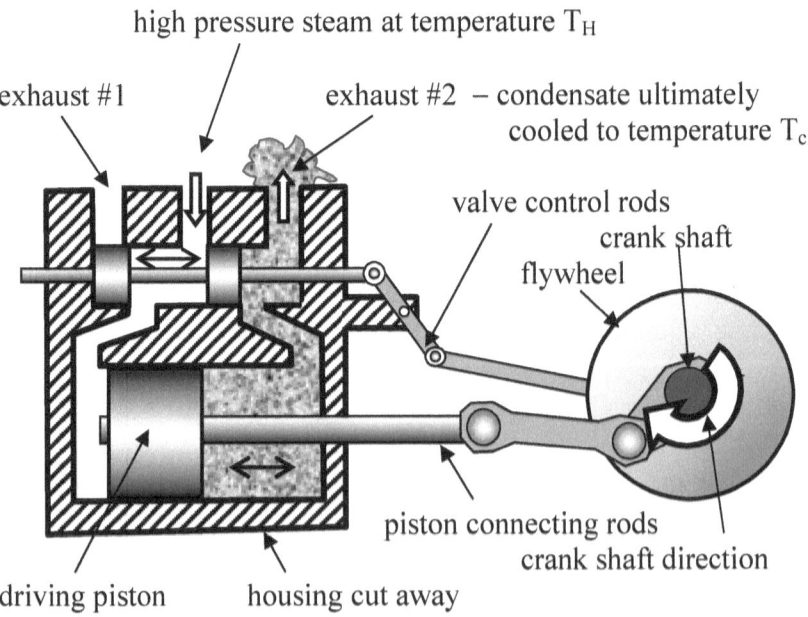

Figure 5.3: Cutaway of a continuous cycle steam engine design

Steam engines in general involve a boiler which is subject to heat generated by the burning of coal, oil, or other material, or even by a nuclear reactor. The boiler produces the steam at a high temperature T_H, which is greatly expanded water vapor. The engine also includes a condenser of some sort that lowers the temperature of the condensate to the cooler temperature T_C.

In the particular design shown in figure 5.3 the heat source, boiler, and condenser are not shown. The boiler is assumed to be connected as the source of the high pressure/temperature steam shown as entering top center. The condenser is, of course, assumed to catch the condensate which exits alternatively at the right or left top of the diagram. The exhaust valve cylinders slide to block one exit during one cycle as high pressure steam forces the main piston to expel the water vapor from the previous cycle out through the other exit. The primary driving piston turns the crank shaft that provides the mechanical work source as well as sliding the exhaust valve cylinder to block the other exit for the return cycle. The design is but one of a multitude that have been implemented.

the Carnot engine cycle

Carnot is best known for discovery of the idealized Carnot cycle[4] whose application is quite independent of specific design. It defined and outlined the process sequences essential to the operation of efficient steam (or other) engines in general to minimize the ubiquitous losses of energy due to irreversibility. This cycle involves the alternating expansion and compression processes by which to continuously convert heat into mechanical work or vice versa. Refer to figure 5.4.

At step 1 the system is in equilibrium, with the high pressure steam at the high temperature T_H. From that situation the expansion reduces the pressure isothermally to step 2 with heat continually being added to maintain the temperature at T_H. At step 2 the exhaust #1 valve opens, further reducing the pressure, but without heat being added so that this expansion is adiabatic. When the temperature drops to T_c at step 3, the

---

[4] Every thermodynamic system exists in a particular state at each instant. A thermodynamic cycle occurs when a system is taken through a series of different states, and finally returned to its initial state as we have already seen. In the process of going through this cycle, the system may perform work on its surroundings, thereby acting as a heat engine that transfers energy from a warm region to a cool region of space, converting some of that energy to mechanical work. The cycle may also be reversed. The 'Carnot cycle' defines the most efficient means possible for converting a given amount of thermal energy into work or, conversely, for using a given amount of work for refrigeration purposes.

momentum of the flywheel forces the piston to compress the gas at Tc isothermally to step 3. Then augmented (in this design) by the alternative steam pressure chamber action on the drive piston and valve openings and closings, the first chamber is returned to step 1 by an adiabatic compression to the temperature of T_H.

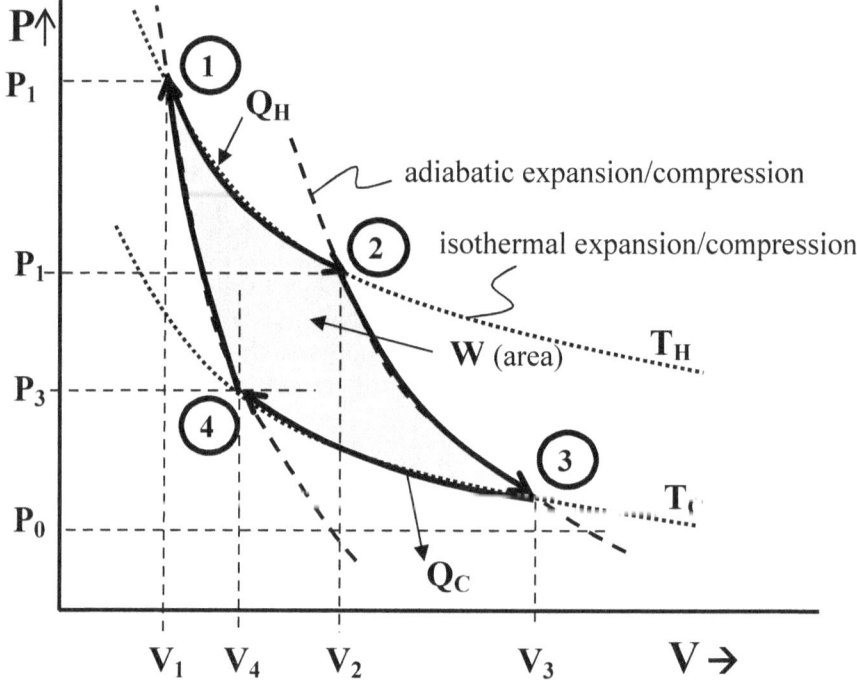

Figure 5.4: The Carnot reversible engine cycle

efficiency expectations

The net amount of work performed during one complete cycle is equal to the integral over the complete cycle of P dV as follows:

$$W_{net} = \int_{①}^{②} P(V)\,dV + \int_{②}^{③} P(V)\,dV - \int_{③}^{④} P(V)\,dV - \int_{④}^{①} P(V)\,dV$$

From the diagram above and our understanding of the conservation of energy, we know that the change in internal energy during one such complete cycle of a Carnot engine is:

$$\Delta U = Q_H + Q_C - W_{net}$$

The net change in internal energy for the complete cycle must be zero since the state of the system is right where it started, so we have that:

$W_{net} = Q_H + Q_C$

Therefore, the net efficiency of a Carno engine cycle must be:

$$\eta_{Carnot} = \left(\frac{W_{net}}{Q_H} \right) = \left(\frac{Q_H - Q_c}{Q_H} \right) = 1 + \left(\frac{-Q_C}{Q_H} \right)$$

The minus sign on Q_C is because it represents the heat that is lost by the system. The ratio of internal heat input to, and output from, the system depends exclusively on the two temperatures T_H and T_C, so that:

$$\eta_{Carnot} = 1 - \frac{T_C}{T_H}$$

So we see that the total amount of heat that is lost in the conversion of heat into mechanical power is dependent on the lower temperature of the environment that takes up the loss:

$$Q_C = - Q_H \frac{T_C}{T_H}$$

Therefore, if the external environment of the engine were at absolute zero, i. e., if $T_C = 0$, one could have a completely 100% efficient engine. But, of course, that is impossible.

That is the lesson that Carnot taught.

some perspective on sources of power

In the last few chapters we have addressed the classical concepts of thermodynamics as they are understood in our mundane everyday world with obvious commercial application. Probably nothing in what we may discover in our continued investigation of irreversibility will change much if any of that from an instrumentalist's perspective. But science is not – or at least the authors feel that it should not be – exclusively about making engines more efficient to thereby become more profitable for corporations.

In the relentless search for sources of power, with ever more power required to propel our civilization, there are increasingly negative side effects. Power plants placed on sizeable rivers raise the temperature of the rivers by several degrees making them less habitable for everything but algae. Our cars, planes, coal power plants, etc. also exude tons of greenhouse gases as exhaust including also sulphur dioxide into the atmosphere. Then there are, of course, the nuclear power plants whose additional waste requires millennia to neutralize.

motivation other than 'motive power'

Scientific investigations should be more essentially epistemological endeavors to enhance understanding of the nature of the universe. What more significant endeavor can there be for mortals such as we are? Should our curiosities not be piqued by what this is to which we have been given such a brief preview? Of course what we learn about the reality of things in such investigations may often be turned into monetary profit as well as improving the animal comforts we enjoy during our time here. But that is not the authors' objective; what we are seeking is an understanding of the underlying causes of the associated phenomena just because we want to *know*. That is reason enough. We see that as in itself the ultimate objective of science. What we, or others, do with that knowledge is an issue not covered by science.

6: Irreversible (*Real*) Processes

"What then do we do about irreversibilities in real processes where the problem can't be solved without paying attention to irreversibilities? You will be horrified to learn that we ignore them! At first, we make believe they aren't there and assume the process is reversible. This allows us to work the problem (incorrectly to be sure), but after we're done we ask how far wrong we are and make corrections to get realistic but approximate answers." [1]

Sadi Carnot's 1824 paper proposed that what he referred to as "motive power", which we just call 'work', results from a drop in temperature from a hot to a cold substance. The second law expresses the fact that over time, these essential differences in temperature, pressure, and density tend to even out in any real physical system that is isolated from an external environment. In any real system differences that might have been exploited initially will ultimately disappear.

How does entropy enter with heat transfer?

So although tweaking what we realize to be impossible reversible processes is the basic approach to applications of thermodynamics just because reversible systems are the only ones for which we can compute answers, we need to understand what all this 'tweaking' is addressing. We discussed reversible thermodynamic processes with either T or P constrained to constant values as we showed in figures 2.5 and 2.6. Implementation involved infinitesimal changes in other primary parameter values with changes made slowly enough to justify the equilibrium constraint of the ideal gas law. That 'law' does not accommodate change, of course, and as such P, T, V and N are the only

---

[1] H. C. Van Ness, *Understanding Thermodynamics*, Dover, New York, 1969 (p. 29)

parameters required in specifying the state of a system in the equilibrium for which it applies. However, when change occurs, whether via a reversible or irreversible process, three more parameters come into play; these are 'work' W, heat 'Q', and entropy 'S', all of which are shown in their appropriate setting in a thermodynamic system in figure 6.1.

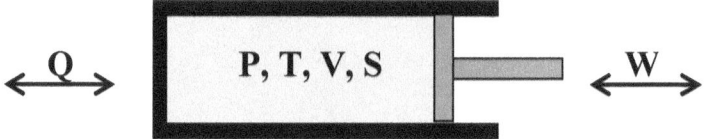

Figure 6.1: Thermodynamic parameters of change

Of course every system possesses internal energy U but it remains constant when there are no changes to the equilibrium status. Similarly, every system possesses an amount of entropy that remains constant for a system in equilibrium. With change come changes to the flow of heat into or out of the system and/or to work performed on or by the system.

For completely reversible processes as shown for the dark curve in figure 6.2, there is no net work done by, or to, the system over the course of a complete cycle and there is no net heat gain or loss. There are, of course, reversible adiabatic changes to temperature and pressure in the process of going between points 1 and 2. So a perpetuated cycle could at least in principle be established in slow motion were it not for the inevitable 'infinitesimal' losses we saw in chapter 2.

Reversible changes in a system maintain a balance in the amount of heat and work that enters and exits a system during a cycle as we have seen in all the cases discussed earlier. The state of a system does not change without a change to either W or Q. In reversible processes, a change in W or Q that takes place in one part of a cycle is undone by a complimentary change in the rest of the cycle. So in proceeding from point 1 to point 2 in figure 6.2 there will be work $+W_{rev}$ performed by the system. That work performed *by* the system will be reversed with work performed *on* the system $(-W_{rev})$ on the return cycle. That will return the system to the state indicated as the starting point 1 in figure 5.2 other than for infinitesimal differences as discussed earlier.

The integral of the work performed between steps of a reversible process equals the change in volume of the gas in that process:

$$\int_{1}^{2} (1/P)\, dW_{rev} = \Delta V$$

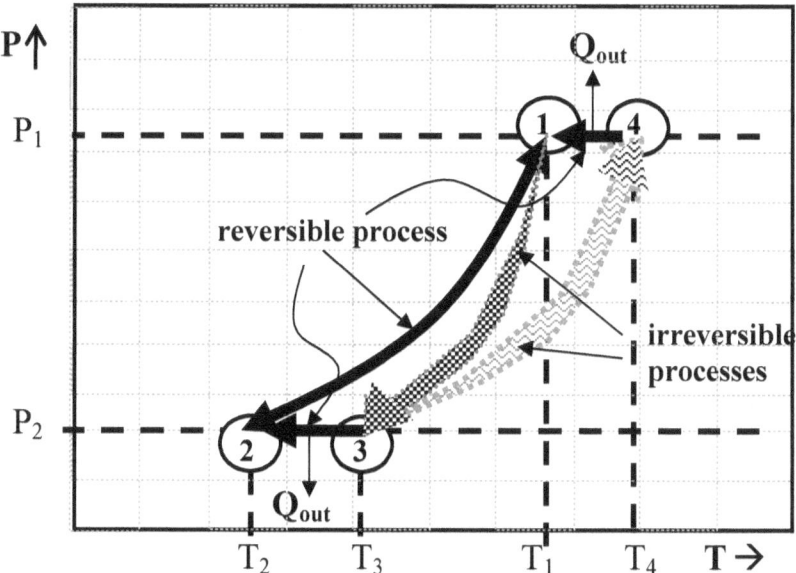

Figure 6.2: Illustration of the comparison between reversible and irreversible processes

This change in volume associated with a reversible change in state between states 1 and 2 is totally independent of the path taken between those states. This implies that the integral above represents an intrinsic 'property' of the system. If we take the integral over a complete cycle we find that:

$$\int_{①}^{②} (1/P)\, dW_{rev} + \int_{②}^{①} (1/P)\, dW_{rev} = 0$$

These facts may not seem all that significant because we knew all along that $dW = P\, dV$ and the net change in all state properties over a reversible cycle would vanish. But what *is* interesting is the analogy that is readily apparent between dW and dQ with regard to changes in the internal energy of systems. If we stress that analogy, we obtain:

$$\int_{①}^{②} (1/T)\, dQ_{rev} = \Delta S$$

which by the same right represents the change in an intrinsic property of the system that is not path dependent for a reversible process. But since there is no change in internal heat during an adiabatic process in going from state 1 to state 2, we have $\Delta Q_{rev} = 0$ and therefore $\Delta S = 0$.

In figure 6.2 we have also shown irreversible process paths, one from state 1 to state 3 that deviates from the reversible path. The precise path of an irreversible process is somewhat indeterminate until stability is established once again as indicated by the obscured width of the path illustrated in figure 6.2. On this path (however it meanders through state to state along its path from state 1 to state 3) an adiabatic process is maintained which does not alter the amount of heat into or out of the system so that we have $\Delta Q_{irr} = 0$ and therefore $\Delta S_{irr} = 0$ on this path as well so that:

$$S_3 = S_1 = S_2$$

However the process has proceeded along this path from state 1 to state 3 the pressure P_2 is attained at the end just as it was in going from state 1 to state 2, but the temperature of the gas will have been raised – whether by friction of the piston on the cylinder wall or by some other phenomenon associated with too rapid changes in the internal state of the gas. It is possible to return the state of the gas from state 3 to state 2, but to do so we must lower the temperature of the gas, which can be done by a reversible process as was illustrated in figure 2.3. To accomplish that, however, we must release internal heat, ΔQ_{rev} as a part of the process, and in doing so we increase entropy as follows:

$$\Delta S = \int_{\text{\textcircled{3}}}^{\text{\textcircled{2}}} (1/T)\, \Delta Q_{rev} = S_2 - S_3$$

Since ΔQ_{rev} is always negative along this reversible path, we have that $S_2 > S_3$ in this case. Therefore since $S_1 = S_2$ after a reversible process, we find that $S_2 > S_1$ after an irreversible process.

How does irreversibility relate to the second law?

Entropy increases in proceeding between any two thermodynamic states whenever the internal heat of the system is changed. Therefore, any complete cycle from, and back to, a given state will always increase entropy if a segment of that cycle involves a change in Q as was shown in figure 6.2. That assures that the process is irreversible.

With irreversible processes, whether there is work put into or produced by the system via changes in volume and pressure there is a continual increase in temperature that precludes indefinite repeatable cyclic behavior as also shown in figure 6.2 for the path segment from state 3 to state 4, etc.. From such enquiries we arrive at the rationale for the second law of thermodynamics which is expressed most succinctly as:

$$\Delta S \geq 0$$

The impact of the second law of thermodynamics on the efficiency of engines, heat pumps, power plants, or indeed any process at all, is profound. We have demonstrated that even reversible processes are not exempt from a loss of total energy in any thermodynamic process. The efficiency of converting heat into power is *always* a proper fraction. So clearly the term 'reversibility' as used in the context of applications of thermodynamics differs somewhat from its use in other branches of physics for which a satellite's orbit would indeed be perpetual if it were orbiting in a vacuum. In these other contexts there is a presumption that there is zero energy loss in whatever process is undergone.

We must admit to a preoccupation with the second of the laws of thermodynamics. It is, after all, this second law that expresses the universality of increasing entropy that has confounded the scientific community for so long. Behavior that seems so clearly reversible in the strictest sense of the term at the submicroscopic level is even more clearly not so in our everyday lives as William James Siddis observed:

> "*And yet, since the reverse universe consists of a perfectly consistent series of positions, obeying all reversible physical laws, it follows that any logical deduction from premises which are reversible laws must inevitably apply to the reverse universe, and that therefore the conclusion must be true in the reverse universe as well as in the real physical universe. That is to say, any deductive conclusion from reversible laws must itself be reversible. And yet, in the case of the second law of thermodynamics, the reversible laws which govern the motions of ultimate particles of matter seem to compound themselves somehow into the best possible example of an irreversible law governing the motions of large masses.*" [1]

The second law states that the entropy of an isolated system that is not in equilibrium will tend to increase over time, approaching a maximum value at equilibrium. Entropy is a measure of the progress of

---

[1] William J. Sidis, "The Paradox," *The Animate and the Inanimate*, Boston: Badger, 1925. This volume is also available on the internet at http://www.sidis.net/ANIM4.

this evening-out process. All versions of the second law have the same effect, which is to describe the phenomenon of irreversibility. The statement concerning the isolation of the system from its environment is significant. Rudolf Clausius first formulated the law as indicating that a process in an isolated system can occur only if it increases the total entropy of that system. Thus entropy takes on extreme significance in thermodynamics.

Reversible processes are exceptions to this rule of the second law of thermodynamics, however. Such processes might include a frictionless 'adiabatic' compression of a substance. But in no case do physical processes decrease the total entropy of an isolated system. Thus, if a system is at equilibrium, by these definitions no spontaneous processes can occur that will drive it away from equilibrium, and therefore the system will remain at its maximum entropy. Any changes must come from outside the system.

A corollary of this law is also due to Clausius, which is that heat cannot spontaneously flow from a substance at lower temperature to one at a higher temperature. From the mathematical definition of entropy, a process for which heat flows in the direction from cold to hot would involve the prohibited decrease in entropy. This is allowable in non-isolated systems of course. But this exception applies only if entropy is created elsewhere, such that the total entropy remains constant or increases as required by this second law. The electrical energy going into a refrigerator is converted into heat that exits the system (refrigerator) in question and results in a net increase in entropy.

Lord Kelvin produced another equivalent form of the second law, which is that it is impossible to convert heat completely into work. This states that it is impossible to extract energy from a high-temperature energy source and convert all that energy into mechanical work. At least some of the energy will be used to heat a lower temperature substance in the process. Thus, an engine with 100% efficiency is a thermodynamic impossibility.

7: Thermal Radiation

"In his lecture Planck stated that according to his rather complicated calculations the paradoxical conclusions of Rayleigh and Jeans could be remedied and the danger avoided if postulates that the energy of electromagnetic waves (including light waves) can exist only in the form of certain discrete packages, or quanta, the energy content of each package being directly proportional to the corresponding frequency."[6]

Thermodynamic systems emit thermal radiation. In fact, anything that is warmer than absolute zero emits radiation characteristic of its temperature. Early experimental analyses of thermal radiation were performed using apparatuses called 'cavities' like that shown in figure 7.1. For obvious reasons what was seen down the hole in the apparatus was called "blackbody radiation". The device is surrounded by a heat bath to maintain the cavity at a fixed temperature. The observed spectrum is basically the surface brightness of the inner wall of the cavity. But that is only because, for usual densities and dimensions, virtually all lines of sight extend to the walls of the container rather than to an intermediate molecule of an associated gas in the container. But in any case, even these are set up to be at the same temperature

The detected radiation invariably has the form given by the Planck radiation distribution. This theoretical form was derived following a notorious failure of its predecessor that was based on classical physics called 'the Rayleigh-Jeans distribution'. Lord Rayleigh's model was based on the equipartition principle from which he derived a dependence of the intensity on the inverse fourth power of wavelength.

---

[6] George Gamow, *Thirty Years that Shook Physics – the Story of Quantum Theory*, Dover, New York, 1966, p. 19.

However, he made a mistake that was later corrected by Sir James Jeans, but even though their analyses were finally impeccable, the predicted form of their radiation distribution, although completely accurate for longer wavelengths, was totally incorrect with regard to its form at short wavelengths, i. e., high energy radiation. This became known as the "ultraviolet catastrophe" because, although it embraced the known laws of physics, it did not correspond with what occurs in nature.

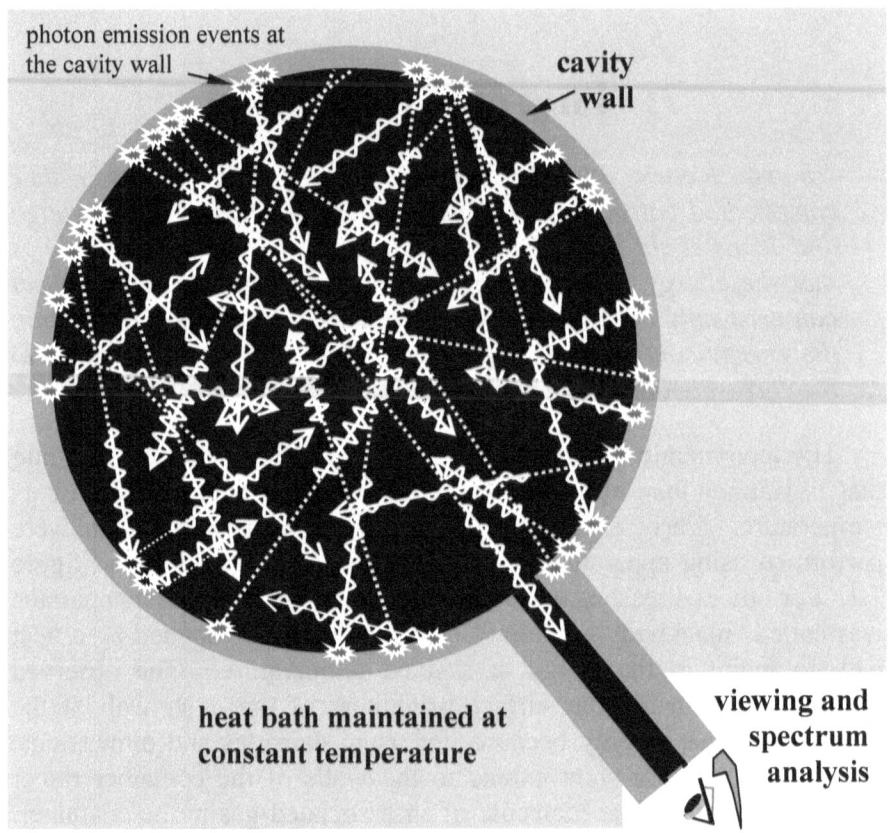

Figure 7.1: Apparatus for observing thermal 'cavity' radiation

Observing radiation from a cylindrical cavity with a slowly sliding piston to assure equilibrium conditions within the cylinder, Wilhelm Wien demonstrated that a leading factor in distributions of blackbody radiation must be a function of the product of the temperature T and radiation wavelength λ so that the total distribution must have the form:

$$\rho(\lambda,T) = f(\lambda \cdot T) / \lambda^5$$

This implies that that the distribution at one temperature has the same general shape as the distribution at any other temperature except that there is a difference of a factor of T^4 as shown in figure 7.2.

characterization of the blackbody spectrum

In thermal equilibrium, the energy density of emitted blackbody radiation will inevitably be distributed by wavelength $\rho(\lambda) \, d\lambda$ with a 'blackbody' spectrum whose functional form is illustrated graphically in figure 7.2. Its formula as it would eventually be determined is:

$$\rho(\lambda, T) \, d\lambda = (2\pi \, h \, c / \lambda^5) \, (e^{K / \lambda T} - 1)^{-1} \, d\lambda$$

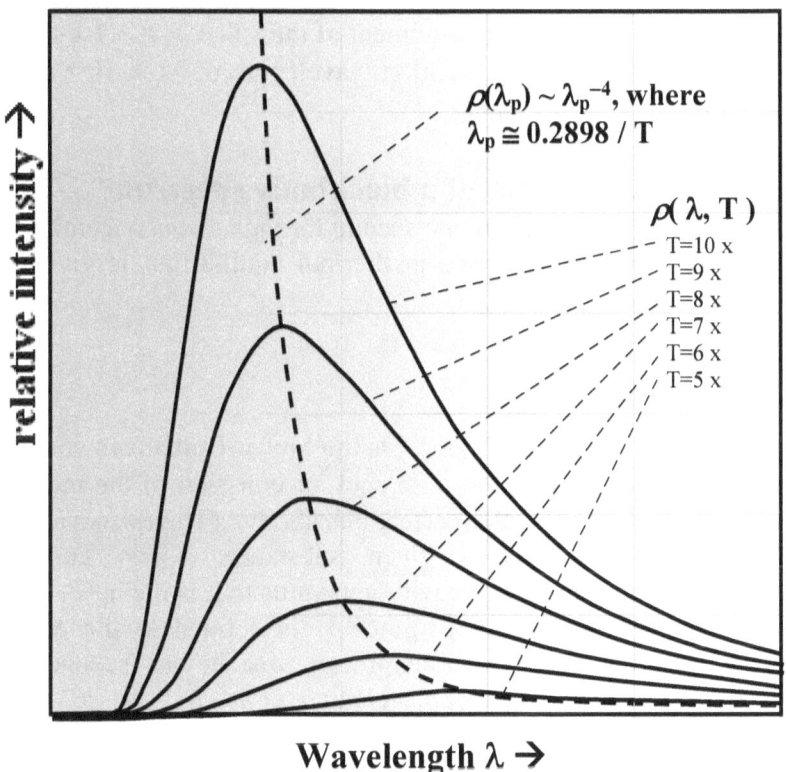

Figure 7.2: The form of the blackbody radiation distribution

This particular parametrical representation is denominated, '*spectral radiant exitance*' and is expressed per unit wavelength. The units of ρ are ergs/cm² sec. The constant factor K in the exponent is given as:

$$K \equiv h \, c / k \approx 1.441 \text{ cm K,}$$

Boltzmann's constant k was defined earlier, c is the speed of light, where c = 2.999 x 10^{10} cm / sec, and h is Planck's constant:

h = 6.626 x 10^{-27} erg sec

It is easily shown that the wavelength associated with the peak of this distribution function is:

λ_p = 0.2 (h c / k T) $\cong$ 0.2898 / T.

When divided by the speed of light, the distribution equation above provides the emission energy per cubic centimeter in the wavelength interval λ to λ + dλ. With the assignment of the value T_e = 3.1 x 10^3 K, this would result in the peak radiation wavelength of λ_p $\approx$ 10^{-4} cm in the infrared domain for example.

temperature dependence of a blackbody spectrum

The total energy radiated in one second through a square centimeter of surface area of any substance in thermal equilibrium is given by Stefan's empirical formula:

$I_T = \sigma \varepsilon \; T_e^4$

Here σ = 2.268 x 10^{-4} erg-cm^{-2}-deg^{-4} is the Stefan-Boltzmann constant and ε is the *emissivity*, i. e., the efficiency of emission of the medium relative to that of a theoretically perfect blackbody. (The *emissivity* and *absorptivity* are equal for a given substance, which is quite understandable with regard to a cavity apparatus that is the appropriate representation of a 'blackbody'.) Since I_T is defined as the energy transported across a one square centimeter area in one second, the radiant energy density in one cubic centimeter is, $E_T = I_T$ /c. And indeed, this is what we obtain by integrating the Planck energy density profile provided above over all wavelengths from λ equal zero to infinity. So the theoretical energy density for a 'blackbody' is:

$E_T = (2 \; \pi^5 \; k^4 / 15 \; h^3 \; c^3) \; T_e^4 \cong 7.56$ x $10^{-15} \; T_e^4$

The curves of overall intensity of blackbody radiation plotted in figure 7.2 involve an extreme dependence on temperature. It is a fourth order effect readily apparent in plots made for several temperatures that show inverted fourth order peak wavelengths, λ_p. A blackbody at twice the number of degrees Kelvin will radiate at sixteen times the intensity as shown in figure 6.2 and solid curves at left in figure 7.3.

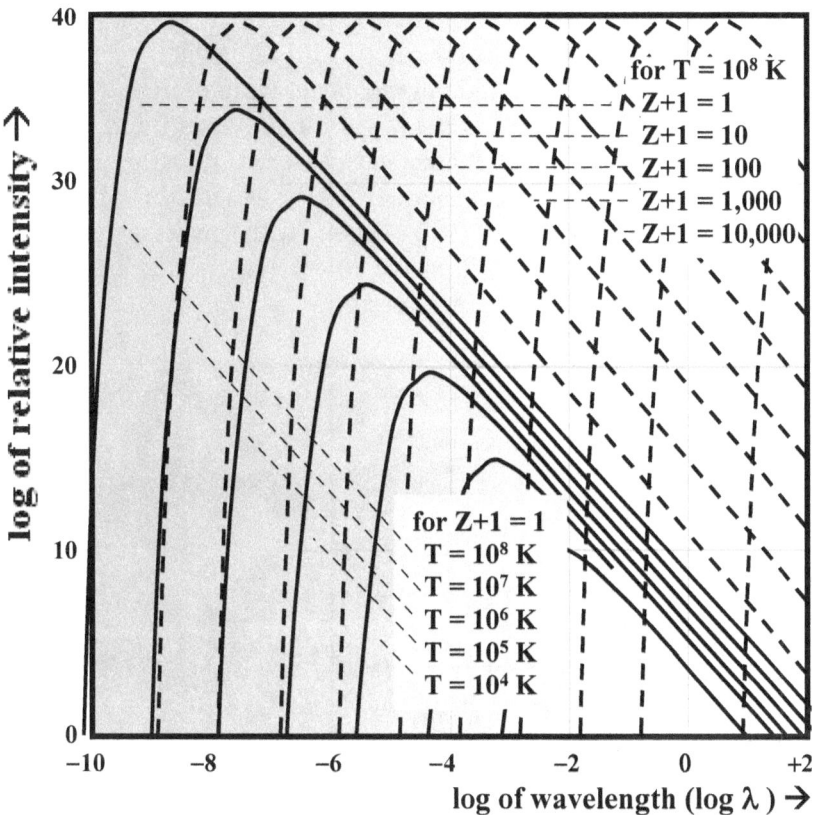

Figure 7.3: The effect of temperature and redshift on spectral distribution

Figure 7.3 illustrates this temperature dependence more clearly on a log scale. As a matter of some interest in cosmology, there is a similar (but different) dependence on redshift Z. This dependence on redshift is illustrated here as having some bearing on the relationship between redshifting of radiation on irreversibility of interactions between radiation and matter. We will have occasion to discuss this effect later.

thermodynamic balance of radiation and matter

In an equilibrium situation determined by conservation of energy the equipartition principle guarantees that the total energy will be partitioned equally among the constituents of a gas. This includes all components of the gas, *not* excluding the photons of electromagnetic radiation.

In figure 7.4 we have depicted the 'heat bath' of figure 7.1 being replaced by an ideal gas of indefinite extent. The gas is assumed to be maintained at the given temperature. Here as in figure 7.1, all emission events occur at the inner surface of the cavity wall. However, for a sufficiently dense and extensive stationary state ideal gas this situation is similar to taking away the solid cavity wall altogether with photons scattering off of particles throughout the gas to similar effect. In figure 7.5 we show only those photons shown in the previous figures 7.1 and 7.4 within the (now merely) conceptual cavity wall. Although most photons originate outside the spherical region where the cavity wall was drawn, interactions all still occur with material particles. Importantly, the distribution within the cavity region will be identical whether the solid cavity wall is there or not.

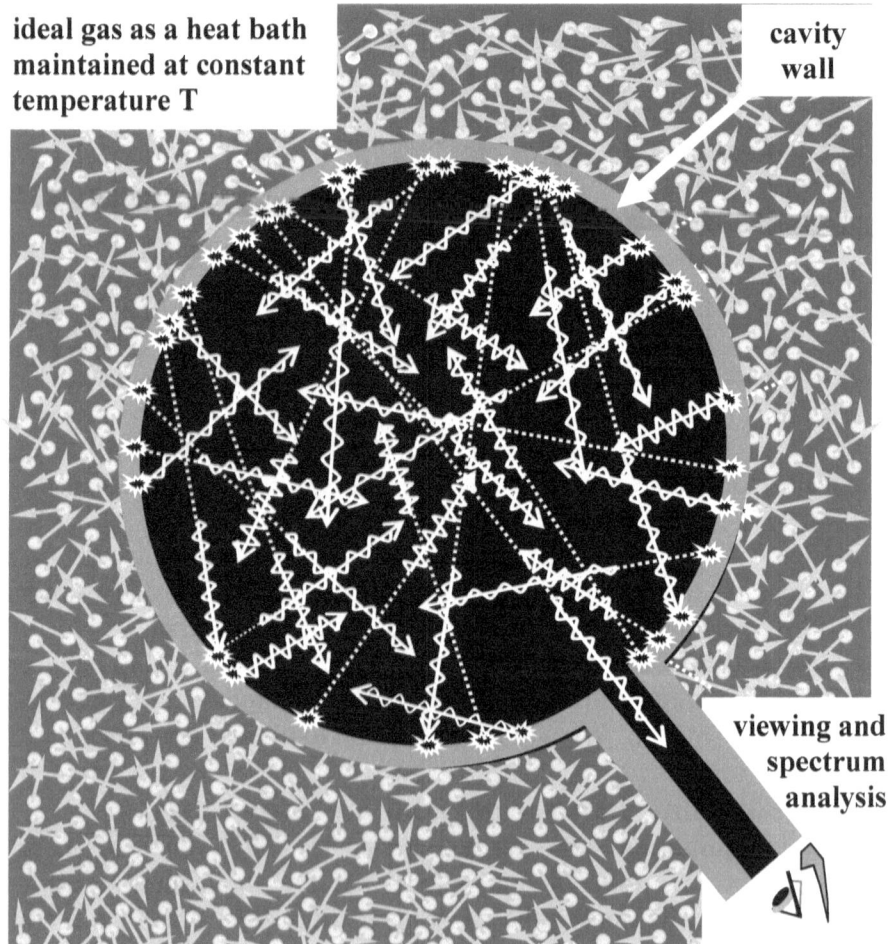

Figure 7.4: 'Cavity' embedded in an ideal gas at a fixed temperature

For cases that involve extensive regions in which exchanges of energy between radiation and matter are dominated by scattering processes, conditions for thermal equilibrium pertain to those regions that are several optical thicknesses interior to surface boundaries. These conditions ensure that radiation and material particles will be in thermodynamic equilibrium throughout such interior regions.

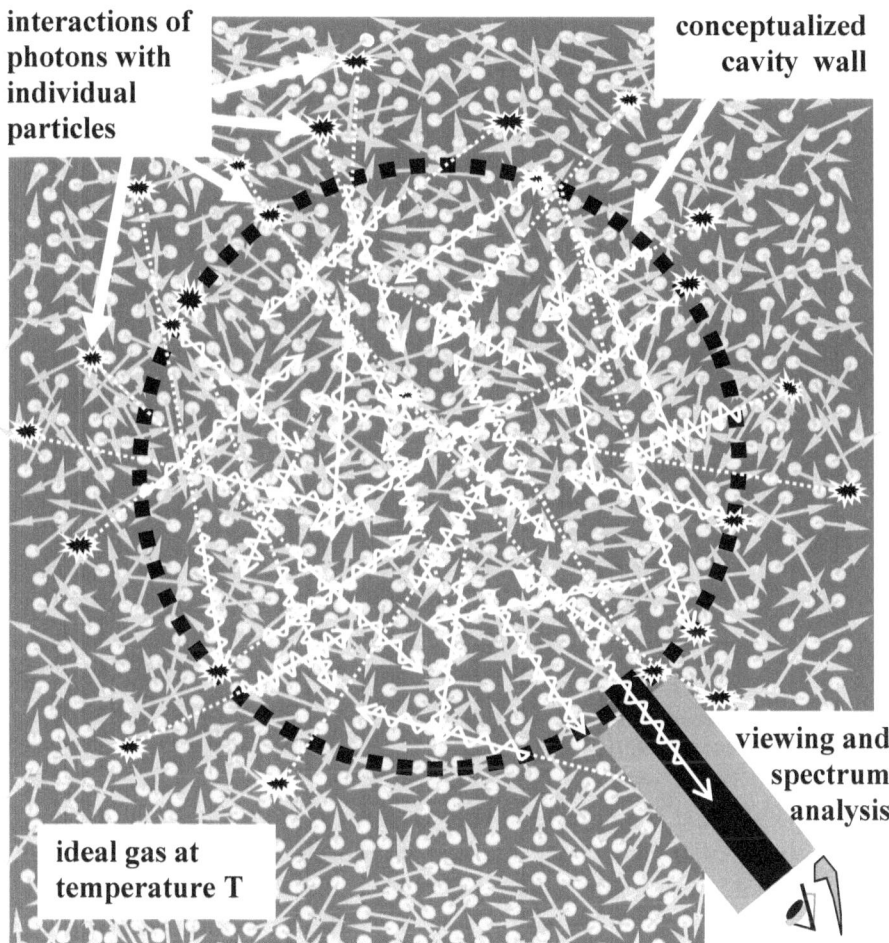

interactions of photons with individual particles

conceptualized cavity wall

viewing and spectrum analysis

ideal gas at temperature T

Figure 7.5: Radiation situation in an ideal gas

In Einstein's quantum theory of radiation (1917) he derived the Planck distribution from first principles using the Boltzmann energy distribution for molecules in an ideal gas. Whereas particles within the equilibrium energy distribution denominated 'Maxwell-Boltzmann' interact among themselves as we will see, photons do not typically (except for rare situations we need not discuss) interact with other photons. So it is only by interacting with matter that photons can be 'thermalized' to their equilibrium Planck blackbody distribution. Other than for thermonuclear reactions, the redistribution of energy in material substances does not alter the number of entities among which the energy will be distributed. This is not the case for electromagnetic energy, however. The number of photons of the various frequencies and the *total* number of photons will definitely be altered by this redistribution process.

Figures 7.1 and 7.5 illustrate two of the ways in which electromagnetic interactions with matter achieve and maintain the blackbody spectrum, either by interacting with molecules and atoms in a solid surface at a fixed temperature or by interacting with particles in an optically thick, gaseous substance that is in equilibrium at the given temperature. These interchanges ordinarily bring about complete energy sharing characterized by the phrase 'equipartition of energy'. It is individual exchanges of energy with particulate matter that result in redistribution of electromagnetic energy to blackbody form. Usually the two instances depicted are extremely similar.

dependence on the density of matter

The blackbody radiation distribution equation does not explicitly depend upon density of the material particles involved in the interactions that bring about thermodynamic equilibrium. This is notably because these equations pertain to surface brightness that incorporates a 'cavity surface' dependence rather than involving volumetric considerations directly. We will eventually address more specifically the way in which surface constraints derive from volumetric considerations appropriate to situations of gases. This will be appropriate to what is illustrated in figure 7.5. But at least in some cases, most notably in cosmology, scattering is involved in maintaining equilibrium of the medium, and when appreciable redshifting occurs there is an additional dramatic effect that results in a difference between the radiation temperature and the kinetic temperature of the medium.

It is indeed significant that the density of material particles does not enter Planck's formula; the reason is that although 'heat' as currently perceived is tantamount to the movement of the constituent particles, the blackbody form of thermal radiation assumes an enclosed cavity as shown in figures 7.1, 7.4 and to the same effect in 7.5. If the conceptual cavity wall shown in figure 7.5 had been less than the optical depth of the medium from the boundary of the substance, then the thermal radiation form would be altered from the blackbody distribution, which is why 'thin' plasmas typically radiate at much lower temperatures than their associated kinetic temperature.

However, there are many implicit relationships between radiation and the temperature of the material substance that do apply. Certainly the ideal gas formula, P V = N R T applies, with P the pressure, V the volume of gas, N the number of moles in the volume, R ≡ N_A k the gas constant, and T the temperature of the gas. This equation is the staple of thermodynamics and must apply to all equilibrium situations. It turns out that it does provide implied gas density dependence:

$$T(\rho_{gas}) = P / R \ \rho_{gas}$$

This equation is plotted as the dotted lines with negative slope in figure 7.6 for various pressure values. These lines correspond to reversible isothermal expansion/compression of an ideal gas.

Here the number density of the gas is defined as $\rho_{gas} \equiv$ N / V. Only by artificially constraining the pressure of an extensive gas could one vary the temperature and density parameters independently with an adiabatic expansion or compression.

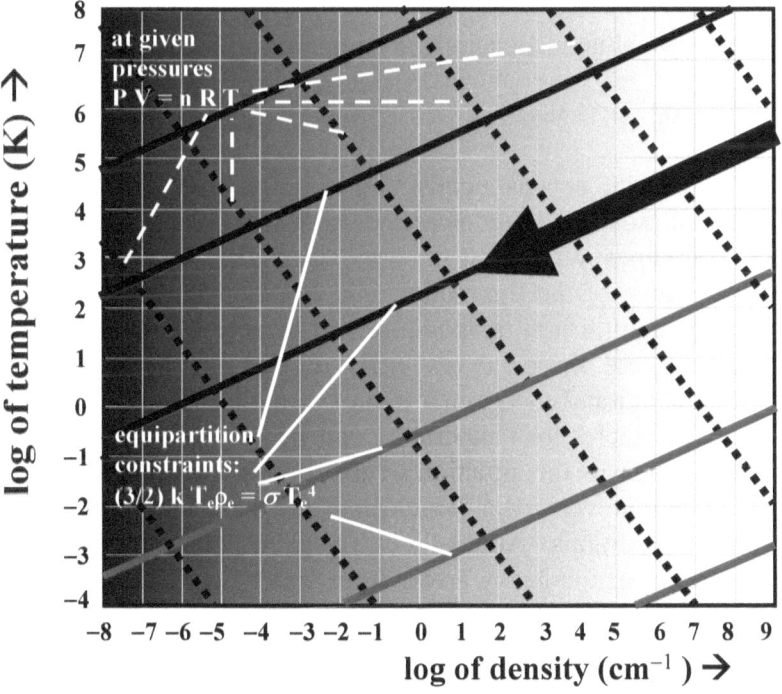

Figure 7.6: **Thermodynamic gas density constraints**

There is another very general temperature-density relationship that applies as well. In particular the energy density of emitted thermal radiation and the kinetic energy density of the gas must be equal at equilibrium according to the equipartition principle. However, if the extent of the gas is such that most lines of sight terminate on a particle exterior to the gas, the temperature of the radiation will reflect that of substances beyond the outer limit of the gas. We will discuss the various possibilities of the radiation temperature emanating from a gas differing from that of the kinetic temperature of the gas in more detail in a later chapter. However, most typically the radiation and kinetic temperature will be in agreement.

The kinetic energy density of a gas in equilibrium at temperature T is given by:

$$E_{\text{Tgas}} = (3/2) \, k \, T \, \rho_{\text{gas}}$$

Thus, in the usual case we will have $E_{\text{Tgas}} = E_{\text{Trad}}$ by the equipartition principle, and we obtain:

$$(3/2) \, k \, T \, \rho_{\text{gas}} = (2 \, \pi^5 \, k^4 / 15 \, h^3 \, c^3) \, T^4 = 7.56 \times 10^{-15} \, T^4$$

From which we obtain a generalized equipartition function:

$$T(\rho_{\text{gas}}) = (\, 0.027 \, \rho_{\text{gas}} \,)^{1/3}$$

This expression acts as a constraint to link temperature and density of matter and radiation in an extended thermal medium; it is illustrated by the solid lines with positive slope in figure 7.6. The constant pressure instance of the thermodynamic equation must give the same results at the equilibrium temperature and density for which it applies naturally. Where these curves cross defines the balance between kinetic and radiational energy in a stable thermodynamic system.

Clearly the scales for temperature and free electron density used in the figure pertain to the situation of microwave background radiation realized in our universe, for which none of these constraints line up with current conditions. Adiabatic expansion espoused by the standard model is shown as the heavy arrow from upper right to lower left with equilibrium maintained until 'decoupling' but not thereafter. We will discuss this in a little more detail later, but the interested reader might want to read a full description in another volume by one of the authors.[7]

---

[7] Ray Bonn, , 2009, pp. 417-509.

thermalization

Thermal blackbody radiation is a phenomenon that cannot be fully understood at the macroscopic level. That is the reason for the nom de plume "ultraviolet catastrophe". Classical physics is totally inadequate to an explanation of the causes of why it is as it is. So unlike the rest of thermodynamics, there could have been no useful theory of what is involved at the level of the immediate observations of the radiation. It is absolutely necessary to look for causes at the submicroscopic level of reality.

It is in the interaction of radiation and matter that the thermal radiation and kinetic temperature come into balance. But to understand that process requires an understanding of the mechanisms of these submicroscopic interactions discussed in the remaining chapters of this book.

Part II

8: Reductionism as Employed in Scientific Enquiry

"I wish we could derive the rest of the phenomena of nature by the same kind of reasoning from mechanical principles; for I am induced by many reasons to suspect that they all may depend upon certain forces by which the particles of bodies… are either mutually impelled to one another and cohere in regular figures, or are repelled and recede from one another; which forces being unknown, philosophers have hitherto attempted the search of nature in vain, but I hope the principles here laid down will have brought some light either to that or to a truer method of philosophy."

– Sir Isaac Newton

Although the current authors are somewhat interested in classical (macroscopic) thermodynamics, and have therefore devoted the seven chapters of Part I to its exposition, it is not the main mission in this volume. The objective here is very narrowly to inspect the nature of individual submicroscopic interactions at the lower level of description that has traditionally been considered completely 'reversible', i. e., as defying the second law of thermodynamics altogether and yet somehow giving rise to the associated phenomena of irreversibility. So although such interactions may not *be* thermodynamics per se, they constitute its most fundamental basis nonetheless. What is it about these tiny systems that supports the notion, or is even compatible with the notion, that larger ensembles of them when studied statistically can be said to result in properties previously unknown within their own domain? That is the question the authors wish to address.

An adequate explanation of the notable disconnect between the explanation of behavior that seems to be completely reversible in laws at work at the submicroscopic lower levels and irreversible at higher levels of these same systems has withheld itself for well over a century. Therefore, the origin of irreversibility that is characteristic of systems generally at the macroscopic level as somehow inexplicable in terms of the more rudimentary basis of thermodynamics at the submicroscopic level of individual submicroscopic interactions is certainly an area where there is an outstanding expectation. It is differences between what happens at these lower levels of our existence and at macroscopic levels for which thermodynamic equations glibly apply that will primarily interest us in the remainder of this volume. But we must fully understand that higher system-level of description in any case so we have addressed that first. Now we proceed with our agenda.

Not content with mere invariant relationships, science has from its inception attempted to understand the more obvious effects by analysis of the less obvious aggregate activities of constituent entities at lower levels. The devil himself is said to dwell among such details, which does not bode well for such undertakings perhaps, although such endeavors have tended generally to meet with much success. This general approach involves the method of reductionism whereby large-scale phenomena are to be explained in terms of lower level (submicroscopic) behavior of constituent elements of the system. This reductionist agenda has proceeded well beyond the intricate atomism of Lucretious and epiphanies of Sir Isaac Newton. Explanations may, and in the case of thermodynamics certainly do, involve in addition to physical properties of submicroscopic particles, the use of probability and statistics.

To address this challenge of unraveling mysteries of irreversibility one must consider how the 'axiom of reducibility' applies in this context. Incompatibility between irreversible thermodynamic processes of the everyday world and reversible processes heretofore conceived as, if not their causes, at least their correlates at the submicroscopic level of the kinetic theory of gases, for example, demands a complete reevaluation. Molecular collision processes must be re-investigated in an attempt to identify heretofore unidentified, but virtually undeniable irreversible effects at that lower level of our existence. Scientific reductionists including the authors can imagine it no other way. 'Emergence' in the context of any scientific discussion is a pseudo-scientific term for 'and then a miracle happens!' However improbable an occurrence may be, be sure that a miracle has not happened; miracles do not happen in valid science.

Despite early enthusiasm and nineteenth and twentieth century successes in achieving the ultimate objective of reducing all apparent complexities that are observed in the physical world to the behavior of submicroscopic particles, currently scientists seem increasingly to back away from that former optimistic view of what is called 'reductionism'. Those who so abjure argue that there must be some sort of 'emergence' of an irreducible *propensity* unprecedented at the submicroscopic level of basic particles. That argument deprecates the earlier scientific world view. It denies the applicability of Whitehead and Russell's *axiom of reducibility* to further inspect what is involved in the most basic physical phenomena of thermodynamics, which has been an approach that has proven of inestimable value to virtually every science. This has only come about because a sufficient counterpart of irreversible macroscopic phenomena has never been identified at submicroscopic levels of our existence. Virtually all of the most basic interactions of particles at this level have seemed, in fact, to be reversible and yet…

Axiom of reducibility – an explanation

The reductionist agenda is to provide an intellectually satisfying explanation of scientific theories, or at least a significant portion of them, in terms of more fundamental concepts operative within the atomic theory of matter. In fact it has proceeded well beyond that level of reality in ferreting out the whys and wherefores of natural phenomena. This transfers the emphasis of analyses from the intrinsic macroscopic properties of a system to a study of the kinematics of the submicroscopic objects of which the system is comprised. Isometric relationships must be established between constructs at the macroscopic and microscopic levels of description to validate this approach.

It has long been recognized that Thermodynamics is an exemplar of reductionism at work. Here, as elsewhere, the axiom of reducibility has proven of inestimable value to understanding invariant physical behavior patterns at the macroscopic level of our everyday existence in terms of the much less immediate submicroscopic behavior patterns. Thermodynamics also, however, demonstrates a dire failure of the approach with regard to understanding the nature of irreversibility.

In order to understand the methodology that embraces a reductionist agenda one must understand the role of establishing an isomorphic relationship between two very different, but intimately related, theories applicable and operative at separate levels of description of the same natural phenomena. This is a relationship whereby theory X at a higher level can be deduced from theory Y at a lower level, once that theory has been firmly established. Three characteristic conditions apply to this methodology as follows:

1) All basic concepts of theory X must be amenable to unambiguous definitions in terms of the concepts of theory Y.

2) All of the natural laws applicable to theory X must be capable of being transformed, by means of the definitions in 1) into laws applicable to theory Y. And these laws must in turn be experimentally compatible with concepts exclusively pertinent to the domain of theory Y. For example, the pressure of a gas is understood to be a higher level descriptive equivalent of the collective individual impulses of molecules of a gas impacting the walls of its constraints.

3) To be useful, the concepts and laws applicable to theory Y must be more *fundamental* than those of theory X. This is in the sense that the impulse of one molecule striking another is much more *basic* and more easily understood conceptually than trying to explain what the 'pressure' of an invisible gas could be otherwise.

Although all concepts of theory X must find a basis in concepts of theory Y, there need be no 1-to-1 correspondence between individual concepts in order to conclude a successful reduction capable of *explaining* the phenomena of theory X. And yes, an adequate rational explanation of observed phenomena is what is desired in any scientific theory.

Clearly entropy and an associated irreversibility defy the first of the three conditions. Should the inability to make an essential connection invalidate the connections between levels that have been found? The answer is clearly, "No." But how should this failure be addressed? It cannot be ignored and to presume that its ultimate failure would not impugn application of the method to understanding the other properties of a thermodynamic system seems ill-conceived. The authors proceed under the assumption that a satisfying connection will be found. That assumption will be amply justified as a conclusion of this investigation.

the relationship between mathematics and physics

This reductionist agenda has been followed intuitively in scientific work for centuries, but became more formalized in the work of Frege, and in Whitehead and Russell's monumental *Principia Mathematica* in which the attempt was made to derive all of higher mathematics from the lower level axiomatic logic and the logical 'theory of types' and 'classes.' A more theoretical formulation of the theory of reductionism is, therefore, to be found in those works where it is treated specifically.

It seems, however, that there is currently more awareness of the failure than of the tremendous success of those efforts in regard to its application to mathematics. A perception persists that their program was unsuccessful because of problems encountered in introducing sets as requisite to considerations of arithmetic and higher mathematics within this same framework. The critical problem in that case was with defining the concept of a set such that it was at once rich enough to support all of mathematics, and at the same time, basic enough to be considered exclusively 'logical'. There is indeed a consensus that they did reconstruct all of mathematics in terms of logic although the feat could only be accomplished by augmenting logic with a concept of sets and auxiliary assumptions constituting axiomatic set theory. Detractors maintain, however, that mathematics is, therefore, *not* reducible to simple logic, although most agree that it can, in fact, be reconstructed as an extension of logic by means of the additional considerations. Of course Gödel's somewhat contrived examples defy rigid resolution, but those issues will certainly not concern us here.

Science, despite more recent efforts to blur the difference, involves a quite different intellectual discipline than mathematics. What scientists have been doing for centuries is associating mathematical statements with the invariances of physical phenomena and then attempting to find some *underlying* principle or generality from which the mathematical statements can be directly deduced. It should be noted that these statements denominated 'laws of nature' are merely descriptive statements, *not* legislated statutes that the universe *must* or *wants to* obey. In this way the laws of nature as codified in mathematical statements have been reduced to a simplified set of assumed universal realities that account for a broad scope of observed behavior. This approach is by its very nature reductionist, so reductionism is both historically and inherently associated with a scientific perspective.

Significantly in classical cases of attempting to reduce the key features of classical thermodynamics (theory X) to constructs and processes in the kinetic theory of gases or statistical mechanics (theory Y), there has arisen one seemingly insurmountable problem with associating theories X and Y that defies isomorphism. This is in addition to the added complexity of there being three levels, X, Y, and Z, i. e., thermodynamics per se, statistical mechanics, and the kinematics of the individual molecular interactions themselves, however one chooses to define levels. Processes defined in the realm of individual interactions at the submicroscopic level of the kinetic theory have traditionally been, and continue to be, understood as inherently 'reversible' whereas those at the higher levels of thermodynamic theory exhibit 'irreversibility'. This difference defies more than just the reductionist agenda; it has defied incontrovertible epistemic explanation of the associated concepts by any approach for well over a century.

Largely due to this failure, there has been an increasing sense of futility precipitated by the perceived impossibility of the reductionist agenda such that rigorous justification employing this method has tended increasingly to be considered by the scientific community as appropriate only for the Don Quixotes of the crank scientific world. Increasingly otherwise-sensible intellectuals tolerate pithy arguments concerning 'emergent' phenomena involving large-scale macroscopic processes conjectured to have no counterparts, or at most only prehistoric vestigial aspects of a 'subsequently evolved' form (whatever that might be taken to mean), at a more fundamental level of reality. The authors admit to being baffled by the prospect of even having to attempt a sympathetic description of such defeatism.

Lest we flatter ourselves with entangled logic, let us acknowledge right off that early man had no problems in distinguishing the differences and similarities between ice and boiling water. Nor did the experiential aspects of heat confuse him. Nor yet was the irreversibility of time something that confounded him any more directly than it does us. But, as one must suspect to be true of most of the evolutionary advantages we have over our progenitors, their limitations in these regards was one of words and symbols in the explanation of the situations they observed. The entire range of modern mathematical tools we enjoy provide primarily a linguistic advantage in the explication of natural phenomena. It allows us to analyze problematic situations and reason more coherently concerning the subtleties of why things are as they are. Our theories are intended to *explain* phenomena.

As we have reiterated frequently, thermodynamics encompasses three separate physical theories at three separate levels of description as well as associated data for which these theories provide explanations. There would ideally be isomorphic relationships between the constructs of each theoretical explanation and the physical phenomena at the next higher level that it in turn explains. This is typical of how physical theories are generally explained and we will achieve that goal.

Be that as it may be, we now proceed to the lower levels of thermodynamic systems.

9: Applying the Axiom of Reducibility to Thermodynamics

"That molecules are endowed with motions follows from the fact that they are assumed to have kinetic energies of agitation and are reflected from walls, producing pressure. That they obey Newton's laws is a consequence of the assumption that they are material bodies in motion, and Newton's laws are assumed applicable to all material bodies." [1]

It seemed evident to Greek philosophers in antiquity that there must be some lower unseen level to reality to which the behavior that we witness traces.   These inventors of the atomic theory included Leucippus and Democritus (both in the 5th century BC) and Lucretius (1st century BC). Conjectures concerning behavior of submicroscopic constituents of matter were contemporaneous with more recent discoveries of thermodynamics.   Avogadro's investigations and quantitative conclusions with regard to the scope and scale of the differences in what we refer to as 'macroscopic' and 'submicroscopic' phenomena were of major significance. Unilateral acceptance of this atomic theory of matter was culminated by Mendeleev's presentation of a periodic table of the elements.

Gradually this lower level atomic theory of matter was elaborated until there is no reasonable doubt about its existence.   Furthermore, there are levels beneath that that are not necessary for understanding the root causes of the validity of the ideal gas law of thermodynamics for example.   Light too, and in fact all electromagnetic emissions that include blackbody thermal radiation have been found to exhibit lower level 'quanta' which are the indivisible basic units of radiation.

---

[1] Leonard Loeb, *Kinetic Theory of Gases*, McGraw-Hill, New York, 1927, p. 15

Isomorphic relationships between phenomena at this lower level and more immediately observed phenomena at the macroscopic level have had to be established in every case. And to accomplish these feats, conjectures had to be made and testing had to be performed in sincere attempts to refute the conjectures in each and every case in the true scientific spirit described so admirably by Karl Popper.[8] This effort has largely been completed by the industrious efforts of generations of scientists. It involves the identification of entities we cannot directly sense but are certain *exist* at the lower level of reality as well as the interactions that *transpire* between and among these entities.

What *exists* at the lower levels of reality?

So let us just elaborate some of the constructs essential to a lower level theory of thermodynamics. In this regard we cite only conjectures that have been overwhelmingly accepted as not having been refuted by experiments designed by the ablest scientific minds. Chief among these conjectures is that all homogeneous matter is comprised of a lowest common unit – that if you continue to split a substance in half, eventually you reach a level at which the remaining substance in each half is a unit that cannot be further split and still have the same substance in both halves. That unit for gaseous substances is a molecule that is the archetype of the substance itself. A molecule is comprised of one or more atomic units of which there are only 98 unique possibilities that are naturally occurring throughout the universe. These atoms are in turn comprised of one of the 98 possible electrically charged nuclei and a number of surrounding negatively charged electrons with various possible energy levels structured in orbital arrangements as described by the 'Bohr atom'. We find it unnecessary to go into the details of these structures; for our purposes molecules are just basic particulate units of the substance. When temperatures become high enough electrons disassociate from the nuclei and form a plasma of charged ions. For our purposes these entities are the lower level objects of which all matter is comprised.

But we have left something out. And that something is radiation. Every macroscopic volume of a substance contains radiation. If that substance is completely encapsulated, then the internal radiation is characteristic of the state of the substance itself. More recent, and the most accurate, unrefuted conjectures involve radiation also being indivisibly particulate in nature. These lowest level units of radiation are photons.

---

[8] Karl Popper, ***Conjectures and Refutations***,

What *transpires* at the lower levels of reality?

There is a categorical difference between what happens to *exist* and what *happens* to those objects that exist. Clear thinking is demanded to avoid categorical errors in this and every other regard.

Clearly the lower level objects that exist in our universe are not static or there would be no associated macroscopic effects for us to observe. These objects interact. There are varying degrees of continuing interaction at the submicroscopic level that characterize static macroscopic thermodynamic states for which a given set of fixed values of temperature, pressure, and volume apply to an equilibrium situation described by the ideal gas law for example.

Those interactions are constrained by what we know of the universe in the way of the conservation laws of nature that describe the facts associated with statements such as "there is no free lunch." Or… if you prefer to have it sung for you, *"Nothing comes from nothing; nothing ever could."* Whenever it has seemed otherwise enquiring minds have gone to work to assure us once again that something from nothing makes no sense. Conservation laws, i. e., conservation of energy and momentum in particular, are the constraints that preclude happenings for which there could be no cause, i. e., something from nothing.

No matter what the interaction, the conservation of energy and momentum must apply to all interactions whether particulate or radiational. What this means is that whatever the energy before an interaction, it must be the same thereafter. The energy of one particle can change but that change must be taken up by another. Also, whatever the net momentum of the system before an interaction, it must be the same thereafter. Otherwise an enclosed system at rest could move off in some random direction because of internal interactions. That cannot happen.

Ultimately, of course, one must develop accurate models of not only the particle interactions and the emission and absorption of radiation separately, but also a composite model of the role of the mutual interaction of particles and radiation. This must include in addition to the reaction impulses caused by the emission out of, and absorption into, associated submicroscopic particles, the scattering of radiation by particulate matter that is part and parcel of what is known as the thermalization process of radiation. But it must also include the radiation that mediates particle interactions. These are major processes that produce the equipartition of energy between particles and radiation, the distributions of particulate kinetic energy and photon energy, and whatever keeps these distributions compatible.

mechanical interactions of molecules

To make sure the reader understands how mechanical interactions between molecules are able to produce the effects observable at the macroscopic level of thermodynamics, let us consider a very simple case of two spherically shaped objects colliding elastically, i. e., with neither object absorbing any of the energy internally upon impact. Suppose we have a spherical object (labeled A) of mass m moving toward a stationary spherical object B of mass M at a velocity v_o directed along the line separating their centers as shown in figure 9.1.

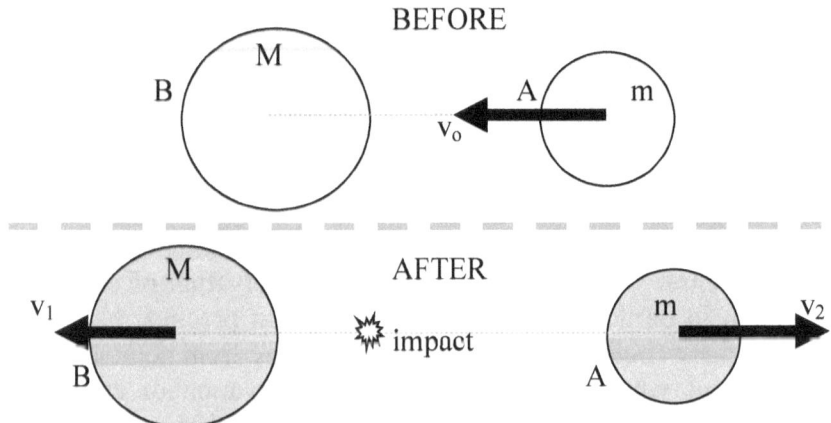

Figure 9.1: Elastic collision of spherical balls with relative motion along their centerline

In this situation we assume we are given the values of M, m, and v_o. The classical problem to be solved is the determination of the values of the two velocities v_1 and v_2 following impact. We assume here, as Boltzmann did, that the laws of classical mechanics apply. Thus, by the conservation of energy before and after the impact we have:

BEFORE → AFTER

$$\tfrac{1}{2} m\, v_o^2 = \tfrac{1}{2} M\, v_1^2 + \tfrac{1}{2} m\, v_2^2$$

So that:

$$v_1^2 = (m/M)\, (v_o^2 - v_2^2)$$

By the conservation of momentum before and after the impact we have:

BEFORE → AFTER

$$m\, v_o = M\, v_1 + m\, v_2$$

So that:

$$v_1 = (m/M)(v_0 - v_2)$$

From which, by squaring the just previous expression for v_1 and setting it equal to the previously obtained expression, we obtain:

$$v_2 = -v_0(1 - m/M)/(1 + m/M)$$

and

$$v_1 = 2v_0(m/M)/(1 + m/M)$$

These two equations are illustrated in figure 9.2.

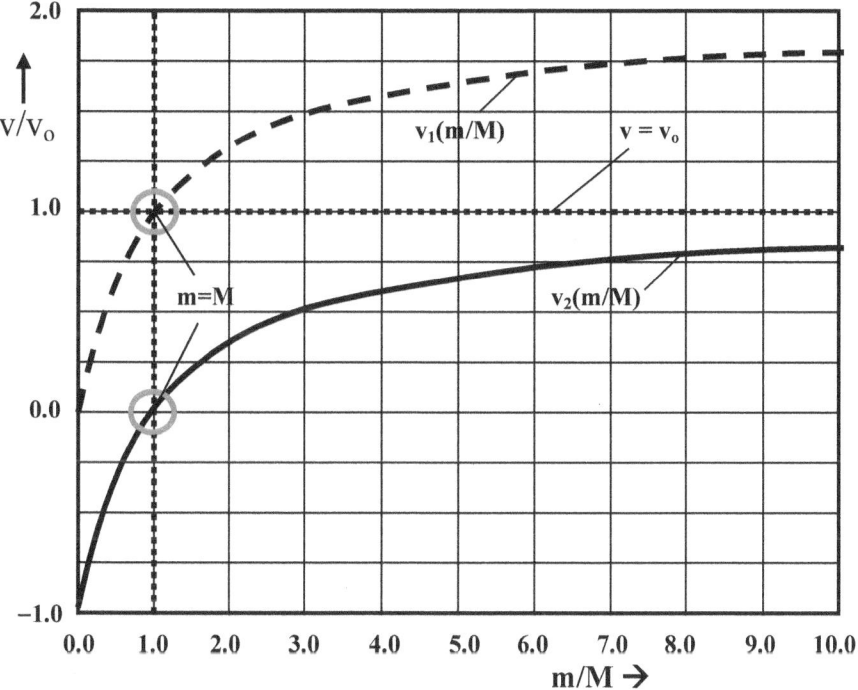

Figure 9.2: Mass dependence of centerline recoil velocities

From these equations it follows that if M is much greater than m (i.e., m/M → 0) then object A will just recoil from object B proceeding in the opposite direction with the more massive object barely affected. If, on the other hand, M is just equal to m (i. e., M = m), then object B will proceed in the same direction as A had been going, with the very

85

same velocity that object A had had, and A will be stopped. If M is smaller than m, (i. e., m/M > 1) then most of the velocity will be transferred to the object B with A proceeding along after it at a slower velocity.

pressure exerted by molecules

Let us apply this analysis in considering the effect of a molecule of a gaseous substance whose mass is m striking the wall of its container. Certainly the wall of the container must be considered an object whose mass M is very much greater than that of a molecule and is therefore effectively immovable. By Newton's laws the change in momentum of the molecule must produce an impulse, I_i, on the wall where:

$I_i = m \, \Delta v_i = 2 \, m \, v_i$

Since force is the amount of the impulse multiplied by the time interval over which it is applied, $F = I \, \Delta t$, we must determine the number of impulses on the wall of the container during a given time interval

For simplicity we will consider a rectangular box container of width A, breadth B, and length C as shown in figure 9.3. If we define v_i as the mean speed of the molecules, then a component analysis of the velocities will yield the equivalent of one third of the molecules impacting perpendicularly to walls in the three directions. A molecule, moving at speed v_i aligned with the length of the container, will in one second travel the length v_i/C times, so it will collide with the two end walls $v_i/2C$ times per second. Since one-third of the molecule velocity components are traveling in this direction, then the number of collisions per second N_{AB} against an end wall of area AB will be:

$N_{AB} = n \, v_i / 6C$

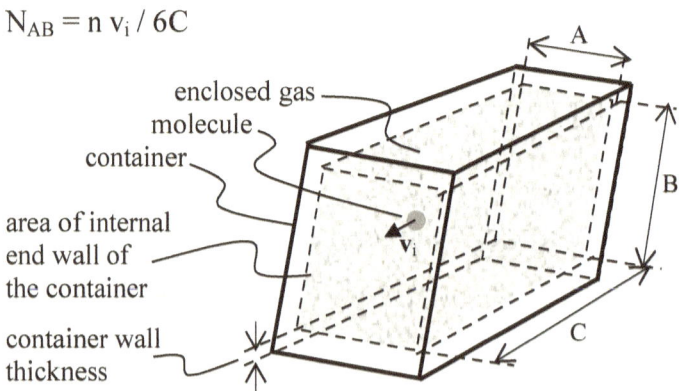

enclosed gas
molecule
container

area of internal
end wall of
the container

container wall
thickness

A

B

v_i

C

Figure 9.3: Fixed rectangular container with enclosed gas

Here n is the total number of molecules in the container, and thus the total impulse of the momentum change of the molecules that is transferred as a force to the end walls in one second is:

$$\mathbf{F_{AB}} = N_{AB}\, \mathbf{I_i} = n\, m\, v_i{}^2 / 3\, C$$

The pressure on the wall due to the combined forces of the impacts is the total force divided by the total areas of the end walls, and thus:

$$P = \mathbf{F_{AB}} / AB = 2/3\, n\, (\, \tfrac{1}{2}\, m\, v_i{}^2\,) / V$$

Here V is the volume equal to the product ABC. Thus, since the parenthetical expression is just the kinetic energy of a single molecule, we have that:

$$P\, V = 2/3\, E_{Total}$$

The conclusions determined here are based on assuming that the size of the molecules is completely negligible and that all molecules exhibit the same speed. The method of the analysis is that which was first conducted by Joule. Clausius performed analyses which took into account the finite size of the molecules, which did not alter the results in most cases. Experiments have found that slight deviations of the inverse proportionality of pressure and temperature do occur in real as against ideal gases. The origin of these variations is indeed in the size, which Clausius considered for larger molecules and in cohesive inter molecular forces.

thermal energy of molecules

Continuing with our analysis, consider that E_{Total} is the sum of the kinetic energy of all of the molecules in the container. Then from the ideal gas law for which PV = NRT, we have that the temperature of a gas is proportional to the total kinetic energy of the molecules. More specifically, we have that:

$$N\, R\, T = 2/3\, E_{Total}$$

Since N in this equation is the number of moles of the gas and N_A is number of molecules per mole, we have:

$$N = n\, N_A$$

So that since $k = R / N_A$, where k is Boltzmann's constant, as discussed earlier,

$$n \, k \, T = 2/3 \, E_{Total}$$

from which it follows that:

$$k \, T = 2/3 < E_i >$$

where $<E_i>$ is the mean kinetic energy of the molecules.

It is easy to see where the conjecture arose that 'heat is nothing more that the kinetic motion of molecules'.

work exerted on and by molecules

To understand the physics behind Joule's discovery of the equivalence of thermal energy and mechanical work, it is instructive to continue discussing the nature of pressure at the submicroscopic level of molecules.

Let us assume that the container with walls of dimensions A, B, and C that we discussed above is equipped with a rectangular piston whose pertinent area is AB and that it can be slid without friction along the direction C. This apparatus is as shown in figure 9.4. The pressure of the enclosed gas would push the piston outward if there were no force applied from the outside. So as we showed above, for the piston to remain stationary there must be a force applied that is equivalent to:

$$F_{AB} = n \, m \, v_i^{\,2} / 3 \, C$$

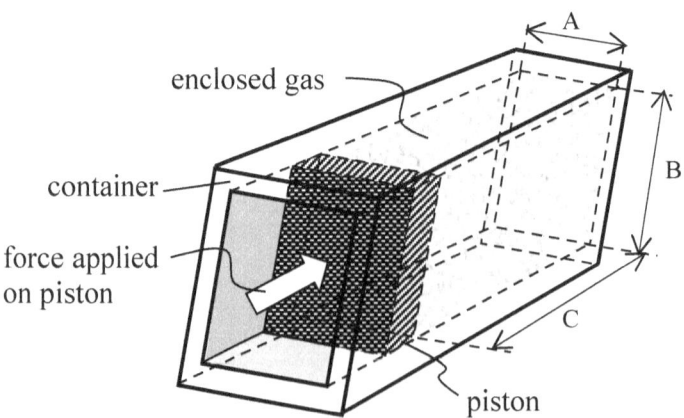

Figure 9.4: Piston arrangement for changing the volume of enclosed gas

If less force is applied the piston will slide outward reducing the gas pressure. If more than this amount of force is applied, it will slide inward compressing the gas. We can compress the gas by changing the amount of force applied to the piston such that it moves inward at the speed v_p. This will change the situation described earlier such that the strength of the impulses produced by the gas molecules striking the piston will become greater than on a stationary wall.

The impact and recoil velocities of molecules striking the piston will now be $v_i + v_p$ relative to the piston, but relative to the fixed portion of the container, the change in velocity will be $v_i + 2 v_p$. Thus, we see that now additional kinetic energy is forced upon each impact of a molecule, with the increase in kinetic energy after each collision equal amounting to:

$$\Delta E_i = \tfrac{1}{2} m \left(v_i + 2 v_p \right)^2 - \tfrac{1}{2} m v_i^2 = 2 m v_i v_p \left(1 + v_p / v_i \right)$$

reversible compression and expansion

Now if, as we suppose, the ratio, v_p / v_i is negligible, we find that the total additional kinetic energy forced upon the molecules of gas in one second is:

$$\Delta E_{Total} / sec = N_{AB} \Delta E_i = n m v_p v_i^2 \left(1 + v_p / v_i \right) / 3 C$$

In the final parenthetical expression we can ignore the final term as long as the speed of the piston is negligible with respect to the mean velocity of the molecules. In this case the amount of energy transferred to the molecules per second by the work expended in pushing the piston is the following:

$$\Delta E_{Total} / sec = P V v_p / C = P AB v_p$$

But notice that v_p is just the rate of change of the length of the container, C, we have that $\Delta V = AB v_p$ and thus the energy imparted to the molecules in a time interval Δt by a slowly moving piston will be:

$$\Delta E_{Total} \Delta t = P \Delta V$$

Therefore, as long as the system equilibrium of the ideal gas law is approximately maintained as it would be by a slowly moving piston:

$$NR \Delta T = P \Delta V$$

And mechanical work and thermal energy can be exchanged as a reversible process, with expansion being treated just as we have compression, but with the sign of v_p reversed.

As we have seen on multiple occasions in earlier chapters and as shown in particular in figures 2.5 and 2.6, in order for thermodynamic processes to be reversible, it is paramount that they be carried out *very* slowly, i. e., $v_p \ll v_i$.

irreversible compression and expansion

So what happens if the ratio, v_p / v_i is *not* negligible? Then the additional term in the parentheses in the previous analysis must be carried along through the calculations, obtaining:

$$\Delta E_{Total} \, \Delta t = P \, (\, 1 + v_p / v_i \,) \, \Delta V$$

So that, significantly,

$$NR \, \Delta T = P \, (\, 1 + v_p / v_i \,) \, \Delta V$$

This imposes a larger change in temperature of the gas during the compression process and a smaller change during expansion with no way to reverse the increase or deficit.

Thus, irreversibility at the macroscopic level of thermodynamic systems can be directly related to the energetics of the molecules at the submicroscopic level. Later we will see how the distribution of the molecular velocities determines the entropy of the system.

the equipartition of molecular energy

Since the ideal gas law is not dependent upon the particulars of the gas involved, if we were to have two identical vessels containing two completely different gases of type A and B, it must follow that,

$$NRT = P_A V_A = P_B V_B$$

and from our discussion of the submicroscopic equivalent of pressure and temperature, we have that:

$$2/3 \, n_A < E_A > \, = 2/3 \, n_B < E_B >$$

By Avogadro's hypothesis $n_A = n_B$ so the mean kinetic energy of the molecules of the two gases must be the same.

This brings us to Dalton's experimental discovery that for a mixture of gases that do not react chemically to each other, the total pressure exerted on the vessel is equivalent to the sum of the pressure that would be exerted by each if it were to occupy the same volume separately. So if we assume the mixing to have been accomplished by allowing gases at the same temperature and pressure to merge from separate vessels into one single container, no kinetic energy would be gained or lost by the diffusion since the collision of particles by which this diffusion process proceeds is always in accordance with the conservation of energy. So for the two (or more) gases,

$$P_A = 2/3 < E_A > / V \text{ and } P_B = 2/3 < E_B > / V$$

Since $< E_A > = < E_B >$, we must have that

$$P_A = P_B$$

limitations of the Newtonian reduction

There is, of course, much more to the subterranean levels of reality than can be accommodated by a simplistic analysis of billiard ball type behavior using Newtonian mechanics. It is amazing, however, that such obtuse phenomena at the macroscopic level can be clarified by considerations involving the simplest version of an atomic theory. We will have occasion to add considerably to the sophistication of our treatment in later chapters.

The mere fact that the methodology precludes a full understanding of entropy suggests that there is indeed much more subtlety to be understood about behavior at the submicroscopic level. Representing this lower level as simply the impact of one molecule upon another totally ignores the role of radiation for one thing. That it has been so glibly assumed that 'heat' associates exclusively with the mechanical motions of particles. That supposed truism is erroneous because it is based on an inadequate representation of the phenomena at the submicroscopic level. It is certainly true that particles with no net charge that are on collision courses will collide directly to effects described above and in more detail in Boltzmann's derivation of the energy/velocity distribution to be described in the next chapters.

But there is a lot of physics that has had to be ignored to obtain such a simplistic analytic approach. Namely, the whole of what has come to be called 'modern' physics has not been called into play in approaches that have been taken. Einstein's relativity and quantum mechanics came into being well after these descriptions of submicroscopic behavior were developed. So we will have to upgrade the analyses to take into account these effects.

But perhaps the major exclusion is the omission of the effects of radiation on the behavior of systems at the submicroscopic level. There are, in addition to reversible direct collisions of molecules that have been admirably addressed, a whole new category of interactions that are mediated by photons of radiation that occur. These interactions are largely irreversible as we will see.

10: Modeling Energy Distributions

"One of the most highly developed skills in contemporary Western civilization is dissection: the split-up of problems into their smallest possible components. We are good at it. So good, we often forget to put the pieces back together again."[3]

The very nature of the Second Law of thermodynamics is in fundamental disagreement with the reversible conservation laws from which it would seem necessarily to derive. But the reductionist agenda that has been a cornerstone of scientific developments for centuries would be in serious jeopardy if deduction of the Second Law from underlying causes were impossible. And yet, how does one even go about modeling such contradictory facets of behavior?

It is incumbent on science to address macroscopic behavior as a consequence of submicroscopic behavior rather than conjuring up the miracles occurring somewhere between by quasi-scientific incantations. So we will have occasion to investigate Boltzmann's methods, both to understand his success and to attempt some corrections for where he failed. We will subsequently attempt to understand the successes and failures of Einstein's derivation of the Planck distribution of radiational energy after first noting the major impact on such models by the concepts of 'modern' physics. We will find that both models presume and exploit reversible processes. We will turn our attention to prevalent heuristic models of entropy that do not take exception to the shortcomings of Boltzmann's and Einstein's models.

---

[3] I. Prigogine and I. Stengers, *Order out of Chaos*, Bantam Books, Toronto, – back cover statement, 1984.

First, however, we must consider the more general nature of open and closed loop models of interactions as a setting for investigating submicroscopic interactions within macroscopic ensembles from two major perspectives. In the next chapters we will reconstruct Boltzmann's analysis of elastic collisions within such a modeling framework. Subsequently we will investigate Einstein's analysis of thermal radiation energy distributions by briefly reconstructing his neoclassical arguments in deriving Planck's distribution of blackbody radiation from his theory of quantum emission and absorption. We will then address non-collision interactions between molecules mediated by electromagnetic photons. This will complete the groundwork from which an irreversible mechanism responsible for the Second Law of thermodynamics can be captured in a more accurate model of the submicroscopic level of reality.

open and closed system models

Scientific analysis requires that systems under study be defined sufficiently that decomposition of an associated model of their behavior into its constituent parts results in no essential aspect being left out and nothing extraneous being added. This discipline would seem to imply that only 'closed' systems could ever be addressed – networks of interacting concepts sufficient unto themselves.

An 'open' loop system on the other hand is one for which there are interactions 'outside' the defined network that are considered beyond the scope of the intended analyses, which are quantified as boundary conditions on the model. So of course open loop models *can* indeed provide a proper framework of analysis if the interactions with external realities are modeled as stylized input and output components interfacing the broader scope of a more global system, not *all* of which is under analysis. Clearly, model preparation is where key differences between reversible and irreversible processes might be expected to slip between the cracks. A system that 'runs down' is obviously a system for which energy is escaping boundaries of the defined 'system'. The Second Law of Thermodynamics challenges the presumption of the feasibility of defining closed systems even at the most fundamental level because in every such system something gets lost. We know that.

The approaches taken by Boltzmann and Einstein have involved conceptualizations that address mutual intuitions that although all systems are essentially *open* at macroscopic levels, strangely at the submicroscopic level they seem to be amenable to complete closure. Boltzmann explored ramifications of a completely closed model of such submicroscopic phenomena. Einstein, on the other hand, identified 'output' to mechanical aspects of the system from his more or less closed model of thermal blackbody radiation, but in so doing ignored the so obvious symmetry requirement of such interactions. Where there is output, there must be a commensurable input if the model is not by its very definition expected to 'run down'.

Boltzmann's and Einstein's modeling approaches are depicted simplistically in figures 10.1 and 10.2. We will eventually address a submicroscopic model that incorporates the processes of both these models in addition to processes ignored by both. The resulting model will be propounded in later chapters of this volume as the most valid portrayal of submicroscopic phenomena underlying all of thermodynamics, including the Second Law in particular.

Clearly, the models depicted below require considerable detail at more refined levels to understand their inner workings, and we will provide that in the next few chapters. We provide this because it is essential to understand the realistic expectation for each of these models based on their scope.

Boltzmann's molecular collision model

Boltzmann modeled the mechanical collision processes as being of paramount significance to the determination of the kinetic energy distribution. He limited his analyses to two disparate types of collision that can occur. These he denominated types I and II as shown in figure 10.1. He considered these two particle collision types as the only means of entry and exit of a particle from a particular velocity interval or restricted domain of phase space. It is, after all, the number of particles within each such partition interval that constitutes the distribution he was seeking.

The two internal circles in figure 10.1 show that any Type I collision will cause a particle to leave the interval [v,v+dv], where v is the velocity of the particle. This is symbolized by path A. Type II collisions can either cause a particle to enter the designated velocity interval symbolized as path B, or remain outside that interval symbolized by the cyclic path C. Boltzmann then introduced a model of each process whose details we will investigate at some length in the next chapter. His model was based exclusively on considerations of the conservation of kinetic energy and momentum using Newton's laws of classical mechanics.

There is obviously a very limited scope to this model, although within the constraints of its assumptions extensibility to a complete

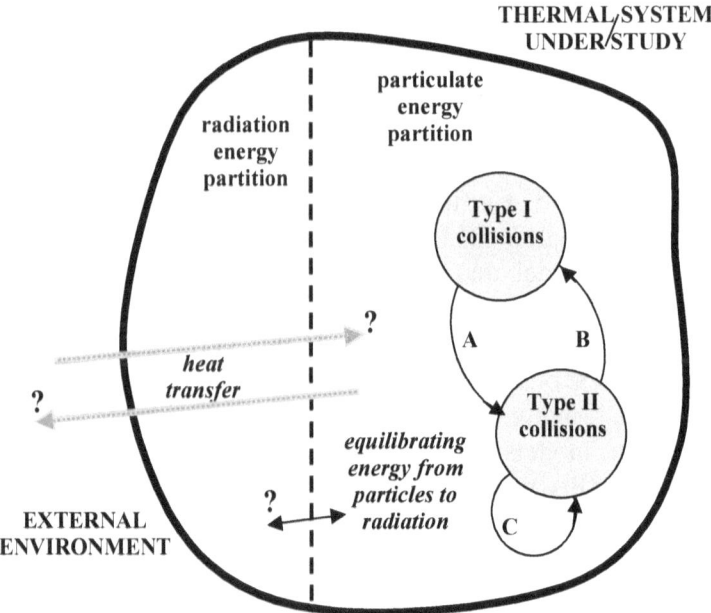

Figure 10.1: Boltzmann's model of collision processes of constituent molecules within a heated substance

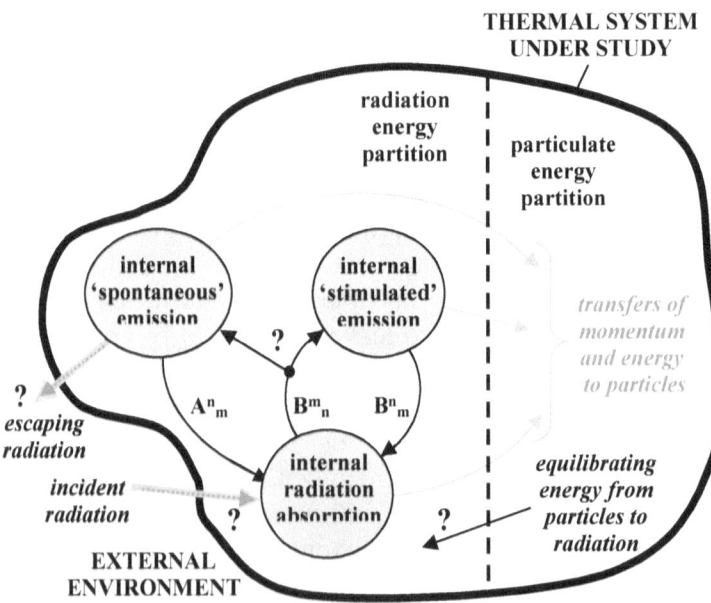

Figure 10.2: Einstein's model of radiation processes of atoms within a thermal 'blackbody'

spectrum of particulate energies is indeed guaranteed. However, he has not considered the possibility of any external energetic interaction such as radiation, affecting the ensemble of particles. This is valid enough as far as exclusively considering particulate material interactions are concerned with neutral molecules on collision courses with each other; the assumption of elastic collisions is generally valid, although those are not the only kinds of interactions that occur between molecules

There is also the major possibility that radiation, whether optical, thermal, or any other kind – none of which was well understood at the time – might in some way distort or augment any of these processes. That molecules exchange radiation that also contributes to changes in velocities of molecules did not seem to have entered Boltzmann's mind. Nor yet is there any consideration of how the equipartitioning of energy between the strictly mechanical and radiational partition might affect the conclusions.

Also, there is the nagging problem of precise synchronization of the entry and exit of remote and entirely unrelated particles to and from the same phase space interval. If a particle leaves the partition before another enters, the distribution of energy would momentarily be altered during that however brief time interval and Boltzmann's equilibrium condition would not precisely apply. Of course the tremendous number of interaction would reduce such time intervals to infinitesimal periods of time. But the degree of this possible variation should probably have been explored more fully.

The extent to which time reversibility of physical processes is equivalent to velocity reversibility is worth consideration here as well. Newtonian mechanics exhibits this velocity reversibility. If all of the particle velocities are reversed, each individual constituent momentum will be directly reversed but the net momentum of the composite of *all* of the particles remains zero. Furthermore, both the individual and composite kinetic energies remain unchanged. Therefore, in an ensemble of constituent particles with zero net momentum such as realized in a contained gas, it should make no essential difference whatsoever to a thermodynamic system whether velocities are all to be reversed or not.

However – and this point is of paramount importance – key features of the equivalence between time and velocity reversibility are lost in going from classical to 'modern' physics, particularly where mediation of interactions between particles is facilitated by quantized energy packets in the form of photons. Radiation emitted and absorbed by constituents in relative motion will also be Doppler shifted in accordance with the relativistic formula. This involves unilateral redshifting precipitated by the transverse components of velocities superimposed upon what is otherwise similar in regard to the classical result for which redshift/blueshift effects would precisely cancel. Classical Newtonian formulas would indicate that there should be no net change in frequency at all in this case.

All this implies that if we take the relativistic quantum realities into account mechanical energy may be lost or gained on each and every exchange associated with mediated interactions between relatively moving constituents for which a transverse component of velocity is involved. Reversing velocities, if the transaction is even viable in reverse, and proceeding as though backward in time will just result in *additional* losses rather than a precise recovery of a previous state. So there is reason to believe that velocity reversibility, and therefore *time* reversibility, may *not* apply at the submicroscopic level of the kinetic theory, if and when, Newtonian mechanics is completely replaced by relativistic quantum mechanics as it must be.

Einstein's radiational model

Boltzmann remained oblivious to the specific effects of radiation since resolution of the appropriate physics still awaited Einstein's genius. Einstein on the other hand was well aware of Boltzmann's earlier work and seems to have considered it to have been virtually complete within its own domain. So again even within an ensemble of atoms or molecules and quantized radiation, not all of the possible loops would be closed.

Einstein addressed the radiation emission and absorption processes of the individual atoms within an ensemble of particles. He argued that there must necessarily be two – each of which he considered unique – emission processes to be modeled. This was a truly ingenious step that allowed him to derive the Planck distribution from first principles that had previously been the mere numerology of an equation that happened to fit the data. Now at last the quantum nature of electromagnetic radiation was incorporated into a thermodynamic model appropriate for study. Refer to figure 10.2.

Einstein also recognized that both the emission and absorption of photons of radiation affect the momentum, and therefore energy distribution of the molecules involved. That is one direction of the previously ignored interaction between the partitions of mechanical and radiational energy accounted for. Resolution of the precise nature of energy and momentum flow from the mechanical energy and momentum of the particles into the radiation intensity distribution was left untouched however.

Despite the ingenuity of the discovery of 'stimulated emission', why there should be two methods for an atom to emit radiation was not addressed at all. Clearly, in the current authors' minds, "God would *not* have done it that way". This is merely an allusion to the sense in which physical theories must satisfy our intellectual expectation of the way things must *be* in order to *make sense* of how the universe operates. It was a favored methodology employed by Einstein. And yes, a scientist must argue for such intelligibility. Simplicity seems to be a primary guideline for understanding all physical phenomena from Occam's razor onward. Multiplying solution alternatives is not the simplest approach to explaining natural phenomena and it is virtually never a part of a thoroughly finalized understanding of that phenomena.

If we consider emission of a photon of radiation from a 'stationary' molecule's perspective, how does one satisfy the conservation laws with the new physics? That problem needed to be addressed and wasn't. And what distinguishes the two types of radiation products in Einstein's analyses? Why would there be two disparate emission alternatives whose products are indistinguishable?

Answers to those questions are essential to closing the loop with the external environment and ultimately to getting a totally closed loop solution that accurately represents reality.

heuristic model of entropy

The study of entropy has evolved into a somewhat independent branch of thermodynamics. Since irreversibility and the concept of entropy are not captured by either the model that describes mechanical motions of particles or the behavior of thermal radiation it has seemed to require its own independent explanation that has increasingly tended to exploit statistical phenomena involving combinatorics more or less deriving from Boltzmann's model. This has given rise to what is called statistical mechanics. This discipline employs the mathematics of probability and statistics as the basis for explaining what are first and foremost physical phenomena. Of course all branches of physics involve mathematical explanation, but in a different sense and degree than its use with regards to explaining entropy.

The second law of thermodynamics tells us that energy always tends to become more uniformly distributed throughout the system, or at least more 'stably' distributed in accordance with the Maxwell-Boltzmann distribution of particulate energies, thereby restricting the amount of useful work it can do. Entropy is seen as providing the measure of that limitation associated with uniformity (i.e., stability) of the energy distribution. In particular it addresses the number of equivalent combinations of submicroscopic particles with given energies at the submicroscopic level that are associated with the macroscopic state of a thermodynamic system. This number increases as a system tends toward an equilibrium situation even though the total energy of the system remains the same.

It is largely velocity v that constitutes the energy of uncharged monatomic molecules in a gas that will be of concern to us here, since classically the kinetic energy of a particle is $E_k = \frac{1}{2} m \mathbf{v}^2$, where m is the mass of the particle. The atomic and molecular particles in a homogeneous gas are all the same, so their rest masses should all be equal except that some molecules will be in a higher energy state due to their having absorbed a photon of radiation. Thus, since it is the square of the velocity that appears in the expression above, it is only the 'speed' of a particle, $v = |\mathbf{v}|$, independent of its direction that affects its energy for such a gas. To the extent that the submicroscopic particles are indivisible, i. e., the system is not at or near a temperature for which particles experience chemical bonding, disassociating reactions, or any thermonuclear forces it is primarily this kinetic energy expression that applies. This simplification also excludes rotational energy, of course. For the more complex cases that we will not directly address here, additional rotational and other energy components would need to be taken into account in the same way as well, but ultimately to the same effect. But even though we ignore those complications, a completely even distribution of energy implying (in our limited case) an even distribution of particle energy would be one for which every particle experienced the very same speed relative to its confines, which is so unlikely for any appreciable number of particles as to be totally inconceivable and furthermore even one grazing particle collision or photon exchange would disturb this unanimity of velocities.

So what is meant instead when we speak of a 'uniform distribution' is one for which velocities (energies) are 'stably' distributed in a 'stationary state', i. e., if a high speed particle is slowed down by an interaction, there must be a slow speed particle somewhere in the system that is sped up by an exchange with another particle to take its place in the overall distribution as demonstrated in Boltzmann's model. Thus, in a 'uniform' distribution, the number of particles with any given energy will not change over time. Boltzmann demonstrated that this kind of stability results for what is called the Maxwell/Boltzmann distribution because the conservation of energy and momentum applies.

In any case according to the accepted view, in order to understand entropy and irreversibility one must understand intricate mathematical aspects of how one assesses the likelihood of a particle possessing a given amount of the total energy of the system. It is that partitioning of the total energy among submicroscopic particles that is considered significant to maintaining stability even as the particles themselves collide and exchange the energy they possess.

To understand the complexity that is being addressed by this approach, consideration is given to the amount of diversity in possible 'placement' of particles into categories associated with the amount of energy allocated to each particle. A particle with no kinetic energy could be allocated to a category labeled '0', for example. A particle with one unit of energy would then be allocated to a category labeled '1', a particle with an energy of two units would be allocated to a category labeled '2', etc.. Although this treatment initially addresses only discrete units of energy 0, 1, 2, etc. times some discrete value ΔE, this analysis is subsequently extended to address a continuous spectrum of energy categories. Then the combinatorics of the allocation is analyzed to illustrate characteristics of the partitioning of energies in a stable distribution.

The approach brings to mind the gyrations of a magician who waves his arms rapidly with incantations during the initial phases of a demonstration designed to distract attention while getting the rabbit into the hat so as to withdraw it in conclusion. The pertinent question to be asked with regard to this approach to modeling behavior in order to explain entropy is, "What is the mechanism whereby the various partitions come to be realized?" As frustrating as it is, the only answer to this question has been, "The reversible collision processes in Boltzmann's model."

11: Modeling Molecular Collisions

"On the one hand, the distribution of velocities in a stationary state is called forth by impacts. On the other hand, the deduction which follows the existence of impacts does not enter at any point, for no use is made of the laws of impact. This proof then, if it be assumed correct, could be used to show that Maxwell's law holds if no impacts took place, a thing which is impossible. A rigorous proof of the law can then only come from a study of the impacts."[3]

We will discuss Boltzmann's profound success in deriving *Maxwell's distribution* of particle energies known to apply at the submicroscopic level of all substances in thermal equilibrium, as well as his notable failure to derive the Second Law using the same approach from the very same conservation laws. Later we will also follow the subsequent unsuccessful efforts to use his probabilistic arguments to circumvent the difficulties encountered. Despite a general acceptance of this latter *complexity* argument even by many prominent physicists on intuitive grounds, all such attempts remain logically flawed. There is a very fundamental mismatch between reversible elastic collisions of microscopic particles involving conservation principles and irreversible behavior at the macroscopic level in the same substance.

collisions as the cause of thermal properties

It was Boltzmann's ingenious recognition that an equilibrium condition arising from collision processes is responsible for the form of

---

[3] Boltzmann's comment with regard to Maxwell's derivation, from L. Loeb, *Kinetic Theory of Gases*, McGraw-Hill, New York, 1927, p. 72

Maxwell's previously demonstrated distribution of energies (velocities) of molecules in gases. Boltzmann derived his *kinetic theory*[3] of gases by assuming the number of molecules that exit any particular velocity region of phase space denominated [v, v+dv] must equal the number entering that region during the same time interval. Thus what he referred to as Type I collisions must be matched by Type II collisions if an equilibrium condition is to exist in the distribution of velocities.

His investigation was based upon the analysis of these two types of collision processes using the conservation laws of kinetic energy and momentum. In his analyses he treated the particles in a gas as though they were mere billiard balls as we did in chapter 9. To provide insight into the analyses that Boltzmann performed, we address the simplest case of two spherically shaped objects colliding elastically, i. e., with neither object absorbing any of the energy internally upon impact as in that earlier chapter. Of course one cannot restrict the analyses to direct centerline impacts as we did in chapter 9 (refer to figures 9.1 and 9.2); one must allow every possible angle of glancing impact of relative velocity to the line joining the centers of the gas molecules at the moment of impact as shown in figure 11.1. Nor, of course, did Boltzmann consider one of the molecules to always be at absolute rest before or after impact; the initial impact velocity is 'relative'.

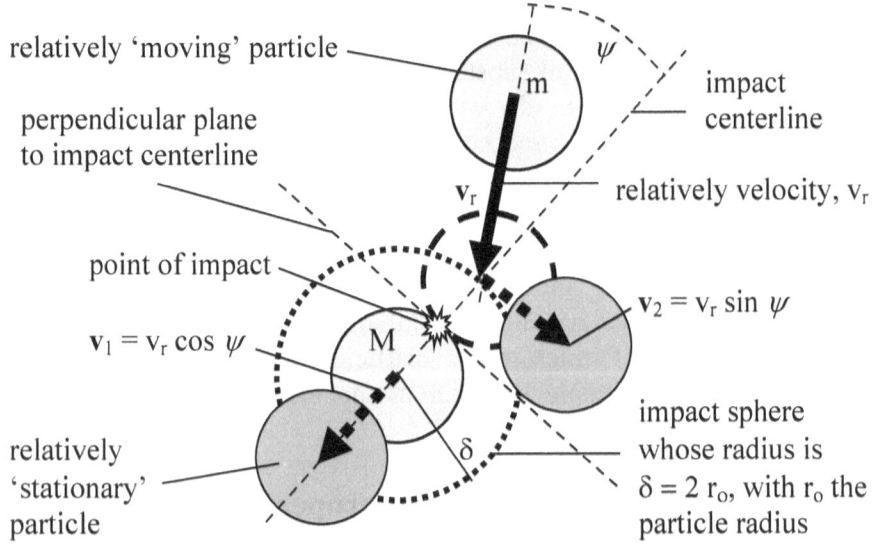

relatively 'moving' particle

perpendicular plane to impact centerline

ψ

impact centerline

relatively velocity, v_r

point of impact

$v_2 = v_r \sin \psi$

$v_1 = v_r \cos \psi$

M

relatively 'stationary' particle

δ

impact sphere whose radius is $\delta = 2 r_o$, with r_o the particle radius

Figure 11.1: Geometry of grazing collisions for m = M

---

[3] Notice however that the particular distribution of particle energies depends upon the type of particles in a system. Fermi-Dirac statistics apply to distributions of electrons and neutrons, for example.

The velocities of the two colliding molecules are significant with regard to the distribution of velocities within the confines of the system, but it is the velocity of one molecule relative to the other that affects the outcome of the conservation equations. So if the velocities of the two molecules are **v** and **v'** relative to the confines of the system, then relative to each other, their relative velocity is:

$$\mathbf{v_r} = \mathbf{v'} - \mathbf{v}$$

as shown at right, with the magnitude being:

$$v_r = |\mathbf{v'} - \mathbf{v}|$$

The relative velocity between colliding particles after the collision will just be $-\mathbf{v_r}$ with an identical magnitude. The only difference between a centerline collision of particles of equal mass and grazing collisions is that in the latter case only the centerline component of the colliding particle's relative velocity is transferred completely to the particle that is struck. The transverse component of the striking particle's velocity is retained by the striking particle. This is the only way that such interactions are compatible with the conservation of momentum before and after the impact shown in figures 11.1 and 11.2.

For a particular collision, the conservation equations for before and after the collision are the following:

	BEFORE
kinetic energy:	$\frac{1}{2} m\, v_r{}^2$
centerline momentum:	$m\, v_r \cos \psi$
tangential momentum:	$m\, v_r \sin \psi$

	AFTER
kinetic energy:	$\frac{1}{2} M\, (\, v_r \cos \psi\,)^2 + \frac{1}{2} m\, (\, v_r \sin \psi\,)^2$
centerline momentum:	$M\, v_r \cos \psi$
tangential momentum:	$m\, v_r \sin \psi$

Solving these equations for any particular case with $M = m$ is, of course, completely straight forward.

range of possible collision situations

It should be noted that there is a plethora of impact scenarios that produce the very same response in the impacted particle, i. e., those for which $v_r \cos \psi$ retain the same value. These variations are shown in figure 11.2 below. The conical arrays of vectors illustrated as centered on the impact centerline for a given impact angle at the top of the illustration all yield the same result as far as the impacted particle is concerned. However, the residual velocity of the impacting particle will be different in every case as shown.

There is in addition a spherical area of possible centerline impacts on the 'impact sphere' as illustrated in figure 11.1. To encompass the total scope of these possibilities, impacts could occur anywhere on this sphere as shown in figure 11.3. This is in addition to the complete range of possible impact angles.

These classical mechanical impact issues are the domain in which Boltzmann chose to address the lower level description of what happens with the interactions of submicroscopic constituents of thermodynamic systems. He did also define a velocity region with regard to which one of the colliding particles could be considered *relatively* 'stationary' with the other particle in relative motion with regard to that region. So defining this quasi stationary region of interest is of paramount importance. But clearly, one cannot take into account every individual interaction and so it is necessary to address the vast distribution of particle situations that involve both impact location and a complete spectrum of all possible individual locations and velocities.

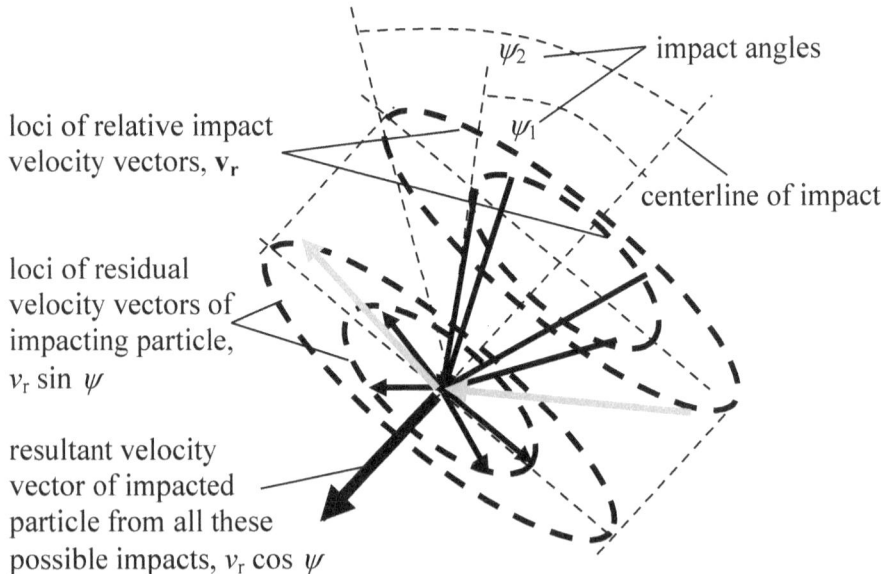

loci of relative impact velocity vectors, $\mathbf{v_r}$

impact angles

ψ_2

ψ_1

centerline of impact

loci of residual velocity vectors of impacting particle, $v_r \sin \psi$

resultant velocity vector of impacted particle from all these possible impacts, $v_r \cos \psi$

Figure 11.2: The many-to-one geometrical relationship between impacting and impacted particle velocities

106

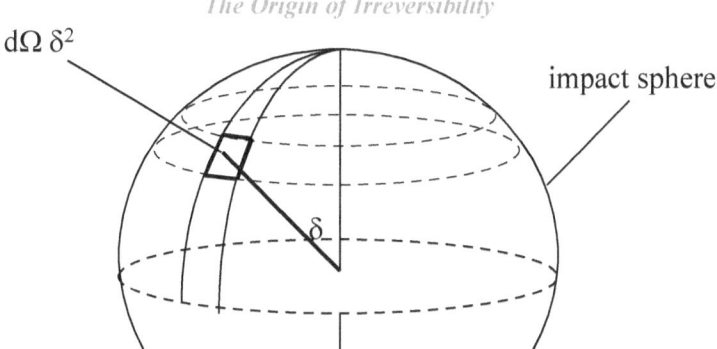

$d\Omega \, \delta^2$

impact sphere

δ

Figure 11.3: Assessing solid angle variations of impact area

spatial distribution of molecules

Boltzmann naturally assumed that every submicroscopic molecule, the total of whose states comprise the macroscopic state of a thermodynamic system, occupies a location within the confines of that system and that it possesses a specific velocity with regard to the position and motion of the confines of the system. Quantum theory alters the validity of those assumptions somewhat, but not much at this level. There are, of course, very many such locations and velocities that must be taken into account. So the distribution of locations and velocities of those molecules must be taken into account.

A spatial sub-region of the system can be characterized by the bracketed pair [**r**, **r**+d**r**] that involves a range of positions whose dimensions are characterized as the three-dimensional differential parameter $d^3\mathbf{r}$ that will be shrunk as one proceeds to the limit of zero size for the region as a part of the integration process. This employs the usual techniques of the integral calculus. Here, of course, we have for the location of **r** the three Cartesian coordinate representation:

$$\mathbf{r} = \hat{i} \, x + \hat{j} \, y + \hat{k} \, z$$

where $\hat{i}$, $\hat{j}$, an $\hat{k}$ are unit vectors along the x, y, and z axes. The three-dimensional differential in this coordinate system is just:

$$d^3\mathbf{r} = dx \, dy \, dz$$

With a natural spherical symmetry this region could also be defined in spherical coordinates about a radial magnitude with two angles that determine the directional aspect of the base position region. Thus, defining r as the magnitude and θ and ϕ as the directional aspects of the position vector **r**, each region's size is given by:

$$d^3\mathbf{r} \equiv r^2 \, dr \, \sin\theta \, d\theta \, d\phi$$

This is typical of the use of spherical coordinates.

In figure 11.4 we illustrate the geometrical aspects that relate the vector position to coordinate parameters to fully represent the location of the sub-region within the system.

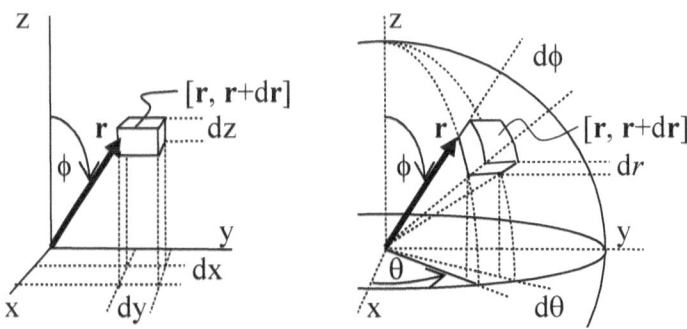

Figure 11.4: Geometrical definitions of the spatial region [r, r+dr]

Because of homogencity considerations of systems in equilibrium, Boltzmann could ignore the spatial aspects of the molecular distribution in his formulation. The probability of a particle occupying any particular place within the confines of the system is the same as for any other place. This is very different than the distribution of velocities of the molecules within the system for which the distribution is far from uniform. The spatial distribution of particles can be expressed as the density of the particles throughout space, $\rho(\mathbf{r})$. In general the total number of particles in a system at a given instant is given by:

$$N_{\text{TOTAL}} = \int^{V_{\text{TOTAL}}} \rho(\mathbf{r}) \, d^3\mathbf{r}$$

Here V_{TOTAL} is the total volume within the thermodynamic system, where, of course:

$$V_{\text{TOTAL}} = \int^{V_{\text{TOTAL}}} d^3\mathbf{r}$$

The homogeneity assumption of molecules throughout the space involves there being the same number within any sub-region of 3-space [r, r+dr] as defined in figure 11.4. This number must remain unaltered throughout time for any system that is in equilibrium since from the ideal gas law, a different pressure and/or temperature would have to be realized for each region for which a different density pertained. This defines a uniform (constant) distribution:

$$\rho(\mathbf{r},t) = \rho = n / V_{TOTAL}$$

Here $\rho(\mathbf{r},t)$ is the density of molecules at a location specified by $\mathbf{r}$ at time t if the distribution were to change over time. However, for an equilibrium situation it will always be a constant independent of time and space throughout a system in equilibrium. Nonetheless, in later developments in statistical mechanics that proceed from where Boltzmann left off, there has been glib consideration of a finite probability of all particles in a system in equilibrium happening to end up in a small subspace of the total volume. For example, all the molecules of air in the room in which you are sitting ending up in one small corner of that room. But, as should be obvious, the conservation of momentum as applied in classical mechanics makes this not just highly improbable, but actually impossible, as we will discuss later.

velocity distribution of molecules

The uniformity and homogeneity assumptions for thermodynamic systems in equilibrium allowed Boltzmann to proceed directly to the more interesting distribution of velocities. This distribution had, of course, already been derived by Maxwell, but there were suspect assumptions that had been made in that derivation.

So we continue here with Boltzmann's more rigorous analytical derivation of that same result. In this analysis it is the sub-region of velocity space that is significant. If a molecule of a gas happens to possess a velocity that places it in the region $[\mathbf{v},\mathbf{v}+d\mathbf{v}]$ illustrated in figure 11.5, it will remain there until and unless it is knocked out of that situation by a collision with another molecule. And if the distribution of particle velocities is to remain unaltered as time passes, another particle must be 'knocked' *into* that velocity situation by another such collision. There are all sorts of collision velocities and angles of impact that can produce this result, as was shown in earlier figures.

We denominate a velocity region of phase space as $[\mathbf{v}, \mathbf{v}+d\mathbf{v}]$ in a directly analogous way to how position was handled for position space above. It involves a range of velocities whose dimensions $d^3\mathbf{v}$ will be shrunk as one proceeds to the limit, employing the usual techniques of the integral calculus. Again employing a natural spherical symmetry, the region is defined as the three dimensional differential interval with a radial magnitude and the two angles to determine any directional character of the base velocity region.

In a similar manner to our discussion of a spatial differential volume, we have **v** is the velocity vector whose magnitude is v. So in Cartesian coordinates the region's size $d^3\mathbf{v}$ is given by:

$$d^3\mathbf{v} = dv_x \, dv_y \, dv_z$$

With the same natural spherical symmetry this region could also be defined in spherical coordinates about a radial magnitude and the two angles that determine the directional aspect of the base velocity region.

$$d^3\mathbf{v} \equiv v^2 \, dv \, \sin\theta \, d\theta \, d\phi$$

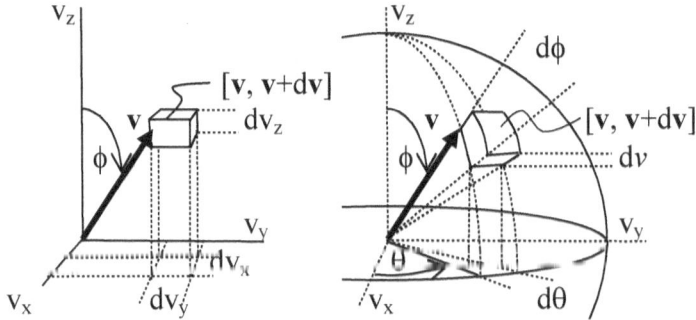

Figure 11.5: Geometrical definitions of the velocity region [v, v+dv]

Here, of course, θ and ϕ are independent of the angles by the same names illustrated for position; they are used her to exhibit the analogy. Again in addition to the magnitude, the angles θ and ϕ specify the velocity's direction in three-space. In figure 11.5 we illustrated the geometry that relates the magnitude with the possible three-dimensional differentials to fully represent the vector quantity **v** and the domain of that portion of velocity space defined by it.

Because of the homogeneity and conservation laws we can depend on the assumption that in every sub-region within the system there will be the same number of molecules with a given velocity as in any other sub-region. Also as we assumed earlier there will be the same number of molecules in each region with a given velocity component in any direction. And therefore, if $f(\mathbf{v})d^3\mathbf{v}$ is the number of molecules with the velocity **v** in the region [**v**, **v+dv**], we must have that the distribution of velocities is coordinate independent so that the overall distribution is just the product of directional component distributions as follows:

$$f(\mathbf{v}) \, d^3\mathbf{v} = f(v_x) \, dv_x \, f(v_y) \, dv_y \, f(v_z) \, dv_z$$

Boltzmann's Type I and Type II collisions

Boltzmann's Type I collisions are those that knock particles out of a particular region of velocity space as illustrated in figure 11.6. According to Boltzmann's logic, if a molecule of the gas does not happen to occupy this velocity region initially, it can only become resident there by reason of having been knocked *into* that region by virtue of having experienced a Type II collision with another particle of the gas as shown in figure 11.7. There are all sorts of these collisions with different relative velocities and angles, because of which we must address all possible situations using functional analyses involving the entire distribution of velocities and impact angles but otherwise as treated above.

Figures 11.6 and 11.7 graphically illustrate the nature of Type I and Type II collisions. A complete description of each collision involves the transition of two particles from their initial to their final situations. The phrase 'relatively moving' refers to motion with regard to the designated velocity region [$\mathbf{v}$,$\mathbf{v}$+d$\mathbf{v}$] of molecule A shown as the box drawn with dashed lines. Similarly, the designation 'stationary' refers to a molecule's occupancy within the referenced velocity region.

In figure 11.6 one molecule is shown as having initially occupied the region [$\mathbf{v}$,$\mathbf{v}$+d$\mathbf{v}$]. It is evicted by the impact of the other molecule, after which neither molecule will occupy the designated region. You will notice that figure 11.7 describes the process whereby a collision between two molecules, neither of which initially occupies [$\mathbf{v}$,$\mathbf{v}$+d$\mathbf{v}$] results in the installation of one of the molecules into that region.

It is significant that what is illustrated in figure 11.6 is a time-reversed version of what is shown in figure 11.7. This isn't just because we chose to draw it that way. We could, of course, have drawn instances of Type I and Type II collisions that were not as precisely time reversals of each other and they would have been legitimate representations of the concept of expulsion and re-entry phenomena. However – and this is very important – non-reversed collision diagrams would obscure the fact (and it *is* a fact) that somewhere else in the system there would have to have been a nearly time-reversed version of each collision at least in the sense of the scope of possibilities shown in figures 11.1 and 11.2 of striking molecule velocities that have the very same effect of evicting the other. It is only *that* class of collisions that is pertinent to maintaining equilibrium. Here is why that is:

In the conservation of energy and momentum equations that pertain to each diagram there is a net change in velocity of the two participants that can only be precisely canceled by a nearly velocity-reversed

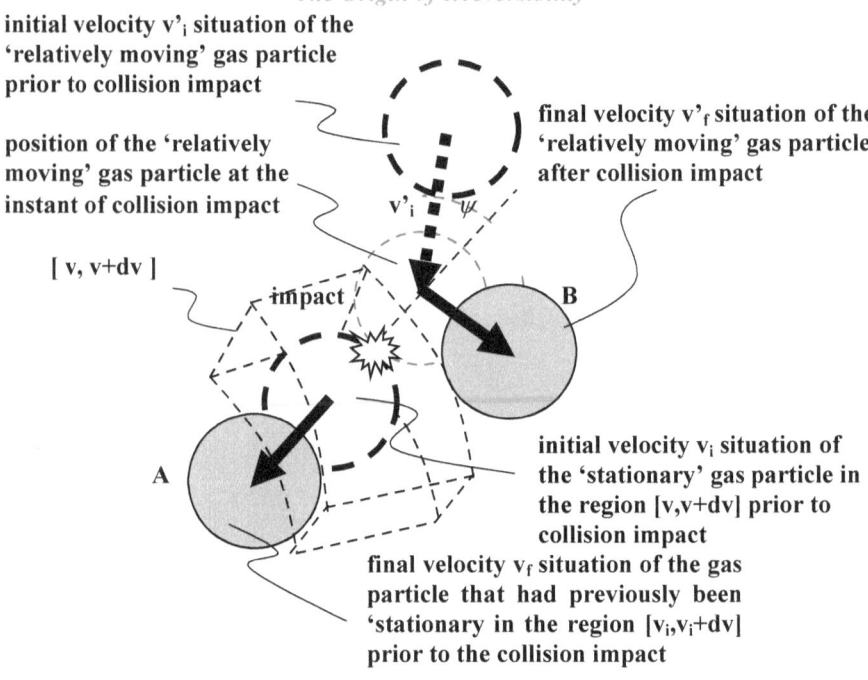

initial velocity v'$_i$ situation of the 'relatively moving' gas particle prior to collision impact

position of the 'relatively moving' gas particle at the instant of collision impact

[v, v+dv]

impact

v'$_i$ ψ

A

B

final velocity v'$_f$ situation of the 'relatively moving' gas particle after collision impact

initial velocity v$_i$ situation of the 'stationary' gas particle in the region [v,v+dv] prior to collision impact

final velocity v$_f$ situation of the gas particle that had previously been 'stationary in the region [v$_i$,v$_i$+dv] prior to the collision impact

Figure 11.6: Type I collisions

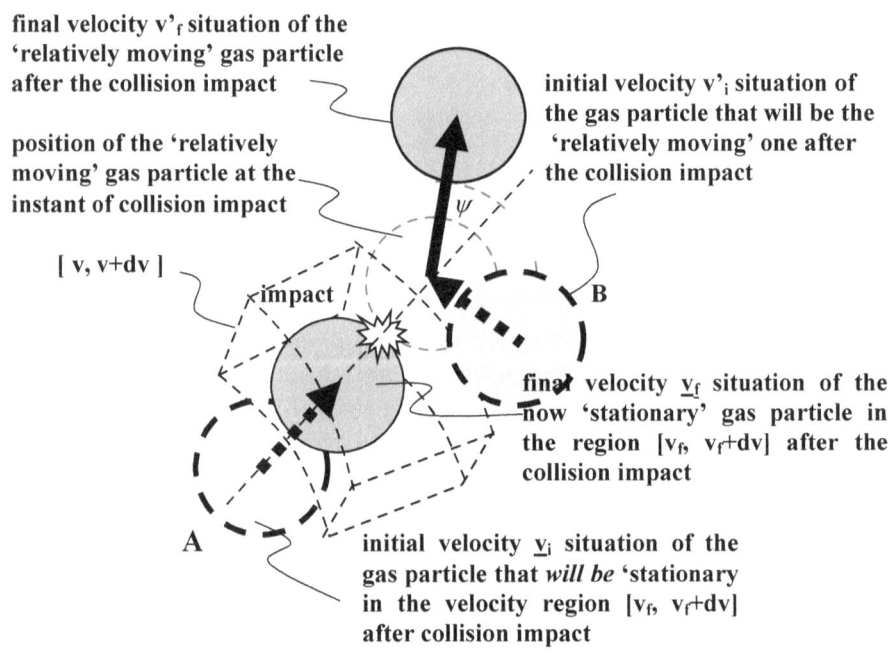

final velocity v'$_f$ situation of the 'relatively moving' gas particle after the collision impact

position of the 'relatively moving' gas particle at the instant of collision impact

[v, v+dv]

impact

ψ

A

B

initial velocity v'$_i$ situation of the gas particle that will be the 'relatively moving' one after the collision impact

final velocity $\underline{v}_f$ situation of the now 'stationary' gas particle in the region [v$_f$, v$_f$+dv] after the collision impact

initial velocity $\underline{v}_i$ situation of the gas particle that *will be* 'stationary in the velocity region [v$_f$, v$_f$+dv] after collision impact

Figure 11.7: Type II collisions

version of that same collision. Somewhere in the system there must be a time reversed counter situation that is a close enough approximation to assure residence within the associated differential velocity interval in order for there to be an unchanging velocity distribution of particles in the system. In figure 11.2 we showed the many-to-one relationship between incoming and outgoing velocities. But for each of the *many* in that relationship there is precise time and velocity reversal.

So Boltzmann's approach as an explanation of irreversibility would seem prima facie to have been somewhat misguided since his argument depends intimately on reversibility. In a later chapter we will amend this situation to show that the photon exchange interactions that were unwittingly omitted from the scope of Boltzmann's model also contribute to the velocity distribution, and these interactions are not typically reversible. But first we proceed with Boltzmann's approach.

The two diagrams in figures 11.6 and 11.7 are the two components of the system model depicted in figure 10.1, the one ejecting a particle (path A in figure 10.1) of velocity v from the region [v,v+dv] and the other inserting another particle (path B in figure 10.1) into that domain. Figure 11.6 depicts the collision of two molecules, one (particle A) in the velocity region [v, v+dv] that will be evicted (to the bottom of the figure) by the collision, thus reducing the number of molecules in that velocity domain by one. Figure 11.7 depicts a collision of two *other* totally unrelated molecules, neither currently in the velocity region [v, v+dv], but one of which will be installed into that velocity interval by the collision along the schematic path labeled B in figure 10.1, thus increasing the number of molecules in that generic interval by one.

Figures 11.6 and 11.7 graphically illustrate the nature of Type I and Type II collisions. A complete description of each collision involves the transition of two particles from their initial to their final situations. The phrase 'relatively moving' again refers to motion with regard to the designated velocity region [**v**,**v**+d**v**]. Similarly, the designation 'stationary' refers to a particle's occupancy within the referenced velocity region.

formulating conservation in terms of velocity distribution

We are dealing with vast numbers of Type I and Type II collisions that require consideration of the density of molecules in such situations, i. e., the number of molecules in the two velocity intervals $d^3\mathbf{v}$ and $d^3\mathbf{v}'$. The uniform density of molecules throughout a thermodynamic system in equilibrium is assumed. However, that is not the case with respect to velocity. The numbers of molecules with the given velocities must be characterized by a velocity density function, $f(\mathbf{v})$, so that the number of situations per cubic centimeter with molecules having velocities $\mathbf{v}$ and $\mathbf{v}'$ in the designated intervals will be given by $f(\mathbf{v})d^3\mathbf{v}$ and $f(\mathbf{v}')d^3\mathbf{v}'$.

Since the molecule labeled A in figure 11.6 that is assumed to be 'relatively stationary' with respect to the motion of the other molecule may actually be moving itself relative to the confines of the system, the distribution of actual velocities of both molecules must be taken into account separately. Thus, the total number of molecules in the Type I situation is increased by the fact of the motion of the base velocity that moves through space during a time interval dt thus increasing the total number of impacts in this class. So to characterize such situations illustrated in figures 11.1, it can be shown that the number of such molecule impacts during a time interval dt will be given by:

$$v \; dt = f(\mathbf{v}) \, f(\mathbf{v}') \; v_r \cos \psi \; d\Omega \; \delta^2 \; dt \; d^3\mathbf{v} \; d^3\mathbf{v}'$$

here $\mathbf{v}$ and $\mathbf{v}'$ are the actual velocities (relative to the confines of the macroscopic system), v_r is the magnitude of their relative velocity, and $d\Omega$ denotes the spherical angular region of impact in addition to variations of the angle ψ as discussed above. The velocity intervals and other parameters have also all been defined previously.

By an analogous symmetry between Type I and Type II collisions it is obvious that the number of molecules reinstated into a region [v, v+dv] must be given by the similar reversed expression:

$$v' \; dt = f(\mathbf{v}) \; f(\mathbf{v}') \; v_r \cos \psi \; d\Omega \; \delta^2 \; d^3\mathbf{v} \; d^3\mathbf{v}' \; dt$$

We cannot immediately assume that the values of the distribution function $f(\mathbf{v})$ and $f(\mathbf{v})$ (the latter distinguished by lack of italics) are identical before and after Type I collisions or that the size of the intervals $d^3\mathbf{v}$ and $d^3\mathbf{v}'$ are the same as those using the same nomenclature in the just previous equation. This latter determination requires some mathematical subtlety to show that equality results because this particular situation qualifies as a special case of problems addressed by the famous theorem attributed to Liouville. Interested readers may pursue that line of enquiry as they wish. We will just accept the result here.

So the total increase Δ_{vt} in molecules in the velocity interval during the time interval dt is given by multiple integration over the range of the involved parameters of the difference:

$\Delta_{vt} = (v' - v) \, dt,$

is as follows:

$$\Delta_{vt} = d^3v \, dt \int_{v'} \int_{\Omega} \{ f(\underline{v}) \, f(\underline{v}') - f(v) \, f(v') \} \, \delta^2 \, v_r \cos \psi \, d\Omega \, d^3v'$$

The number of gas particles involved in the region d^3v including the effects of both Type I and Type II collisions is, of course, just the sum total of the numbers of molecules expelled from and added to the region during the time interval dt. The extent to which this number is stable is determined exclusively by the balance between Type I and Type II collisions. It is therefore only this difference that drives the velocity distribution of a gas to equilibrium. Furthermore, once a stable distribution is realized, it can only be the equivalence of the number of Type I and Type II collisions occurring in each and every region during every time interval dt that is responsible for maintaining that stable equilibrium distribution. So although there are always changes to the velocities of particular molecules, the flow of molecules into and out of every velocity region must always be the same.

In a macroscopic equilibrium situation for which the lower level distribution of submicroscopic particle energies (velocities in this case) remains unchanged, the numbers of the two respective types of molecular flows within the single velocity cell of phase space must remain equal for each and every such velocity interval during every infinitesimal time interval dt. Notice that most Type II collisions in the system will not directly affect the number of molecules in any one particular interval [v,v+dv] since the impacts will be too hard or too soft, the wrong angle, etc., and therefore, of course, the schematic path C in figure 10.1 will most often be taken. However, each of these collision situations will be pertinent to *another* such interval although for a different infinitesimal velocity region. Many of Boltzmann's significant contributions to thermodynamics derive from just this simple equilibration condition on paths A and B in his model shown in figure 10.1. The approach can be directly extended to include other modes of mechanical energy by taking into account additional conservation laws associated with those degrees of freedom.

From the definition of the velocity distribution functions $f(v) \, d^3v$ we have that the net change in the number of molecules in the given velocity region must be given by the rate of change formula:

$\partial/\partial t \, (f(\mathbf{v}) \, d^3\mathbf{v}) = \partial f(v)/\partial t \, d^3\mathbf{v},$

since the region size does not change. Equating these two expressions,

$$\partial f(\mathbf{v})/\partial t = \int_v\int_\Omega \{\ f(\underline{\mathbf{v}})\, f(\underline{\mathbf{v}}') - f(\mathbf{v})\, f(\mathbf{v}')\ \}\ \delta^2\,|(v'-v)|\cos\psi\,dv'\,d\Omega$$

And since the net change in the number of molecules in each velocity region must vanish, we find that the condition:

$$f(\underline{\mathbf{v}})\, f(\underline{\mathbf{v}}') - f(\mathbf{v})\, f(\mathbf{v}') = 0$$

would be sufficient to guarantee stability. It is *sufficient* to force the net change in the distribution due to all possible collisions to zero and the molecular velocities would therefore all be in equilibrium. So if it could be shown that this difference always has the same sign, then it could be concluded that the condition is not only sufficient, but also a *necessary* condition for equilibrium. This is because there are possible modes whereby fluctuating values of an integrand can result in a zero integral without requiring that the integrand itself be identically zero.

The relationship between a sufficient and a necessary condition for this particular situation is illustrated in figure 11.8.

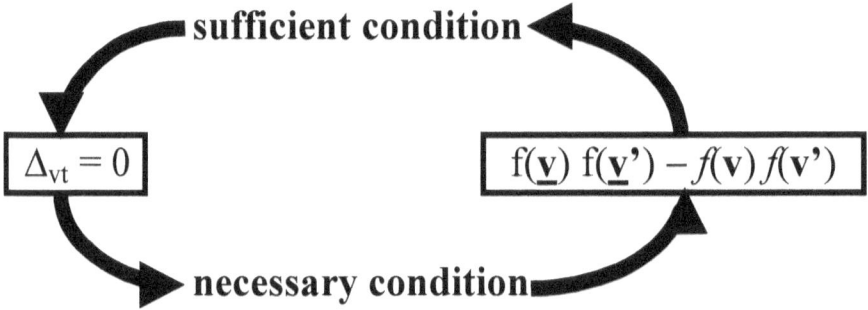

Figure 11.8: Necessary and sufficient conditions for a stable distribution

Boltzmann's 'H Theorem'

Much of Boltzmann's career, and the celebration thereof, involves this subtlety of expanding the scope of his conclusion from that of a *sufficient* condition to a *necessary and sufficient* condition. For this he defined what he called the '*H* function' that is exclusively dependent on the *form* of the function $f(v)$ as follows:

$$H \equiv \int_{\mathbf{v}} f(\underline{\mathbf{v}}) \log f(\underline{\mathbf{v}})\, d^3\mathbf{v}$$

Since the distribution $f(\underline{v})$ is defined as the equilibrium distribution, it does not vary with time so that differentiation of H with respect to time at equilibrium must yield:

$dH/dt = 0$.

But just differentiating this seemingly arbitrarily defined function by time obtains:

$$dH/dt = \int_v \partial f(v)/\partial t \, (1 + \log f(\underline{v})) \, d^3\mathbf{v}$$

Substituting for $\partial f(v)/\partial t$ from the previous equation one obtains:

$$dH/dt = \int_v \int_{v'} \int_\Omega (1 + \log (f(\underline{v}))) (f(\underline{v}) \, f(\underline{v'}) - f(v) f(v'))$$
$$\delta^2|(v'-v)| \cos \psi \; d^3\mathbf{v}' \, d^3\mathbf{v} \, d\Omega$$

The same procedure must be followed for the population of the region $[\underline{v}', \underline{v}'+d\underline{v}']$ characterized by $d^3\mathbf{v}'$. Then by subtracting, one obtains:

$$dH/dt = \tfrac{1}{4} \int_v \int_{v'} \int_\Omega \log (f(\underline{v}) f(\underline{v'}) / f(v) f(v')) (f(\underline{v}) f(\underline{v'}) - f(v) f(v'))$$
$$\delta^2|(v'-v)| \cos \psi \; d^3\mathbf{v}' \, d^3\mathbf{v} \, d\Omega$$

Since ψ is always less than $\pi/2$, the integrand will always be negative because the two bracketed quantities necessarily have opposite signs. We find that at equilibrium the following condition is indeed *necessary*.

$f(\underline{v}) \, f(\underline{v'}) - f(v) f(v') = 0$

So far we have treated the functions $f(\mathbf{v})$ and $f(\mathbf{v'})$ as velocity 'densities' (i. e., as the numbers of molecules of a given velocity per unit volume) analogous to density of molecules $\rho(\mathbf{r})$ in 3-space. Since the molecular density in 3-space is uniform and therefore constant, we can divide the f-functions by the uniform molecule density in space and thereby obtain probability distributions. At this point we can, therefore, make the change in definitions without altering the expressions in the previous equation in any way. And the previous condition makes it possible to determine the form of the distribution function $f(v)$.

solving for the velocity distribution

Although the velocities in Type I and Type II collisions are similar to the reversed set, they are not exactly the same. The functional values differ because the specific velocities differ, which matters when they are not merely 'dummy' values within an integration step. But we do have that by the law of the conservation of energy the following equation must hold:

$$\underline{v}^2 + \underline{v'}^2 = v^2 + v'^2$$

The logarithmic form of the just previous necessary condition equation on the distribution function becomes,

$$\log f(\underline{v}) + \log f(\underline{v'}) = \log f(v) + \log f(v').$$

Since isotropic substances that we are assuming pertain to our thermodynamic system cannot have distributions with velocity sign dependence the direction of molecule velocities can be of no particular consequence. Thus, it can only be the magnitudes of those velocities that affect the form of the distribution. And we therefore can conjecture as values for the four distribution functions, the following:

$$f(v) = e^{\sigma(v^2)}, \; f(v') = e^{\sigma(v'^2)}, \; f(\underline{v}) = e^{\sigma(\underline{v}^2)}, \; f(\underline{v'}) = e^{\sigma(\underline{v'}^2)}$$

Thus, from the logarithmic equation above, we arrive at,

$$\sigma(v^2) + \sigma(v'^2) = \sigma(\underline{v'}^2) + \sigma(v^2 + v'^2 - \underline{v'}^2)$$

by substituting

$$\underline{v}^2 = v^2 + v'^2 - \underline{v'}^2$$

using the conservation law equation. Then by successive partial differentiation with respect to v^2, v'^2, and $\underline{v'}^2$ it is easy to show that,

$$\partial\sigma(v^2)/\partial v^2 = \partial\sigma(v^2)/\partial v^2 \, (v^2 + v'^2 - \underline{v'}^2)$$

$$\partial\sigma(v'^2)/\partial v'^2 = \partial\sigma(v'^2)/\partial v'^2 \, (v^2 + v'^2 - \underline{v'}^2)$$

$$\partial\sigma(\underline{v'}^2)/\partial\underline{v'}^2 = \partial\sigma(\underline{v'}^2)/\partial\underline{v'}^2 \, (v^2 + v'^2 - \underline{v'}^2)$$

and thus that,

$$\partial\sigma(v^2)/\partial v^2 = \partial\sigma(v'^2)/\partial v'^2 = \partial\sigma(\underline{v'}^2)/\partial\underline{v'}^2,$$

This can only be true if each of the partial derivatives is equal to a constant. We choose it to be $-1/\alpha^2$ for all three of the partial differentiation cases. And therefore, by integrating and defining the log of A^3 as the constant of integration, we obtain:

$$\sigma(v^2) = -v^2/\alpha^2 + \log A^3$$

Thus we arrive at a formulation of the probability distribution of velocities of molecules in a monatomic gas:

$$f(v)\, d^3v = e^{\sigma(v^2)}\, d^3v = A^3\, e^{-v^2/\alpha^2}\, d^3v$$

And because $f(v)\, d^3v = f(v_x)\, dv_x\, f(v_y)\, dv_y\, f(v_z)\, dv_z$ as shown above,

$$f(v_x)\, dv_x = A\, e^{-v_x^2/\alpha^2}\, dv_x$$

for all components of molecular velocities.

The reader may be dismayed to discover that after all this arduous derivation, the entire scheme will require some re-evaluation in lieu of discoveries made after Boltzmann had completed his analyses.

12: Properties of the Maxwell-Boltzmann Distribution

"Maxwell demonstrated that this particular state, which is the thermodynamic equilibrium state, occurs when the velocity distribution becomes the well-known 'bell-shaped curve,' the 'gaussian,' which Quetelet, the founder of 'social physics,' had considered to be the very expression of randomness. Maxwell's theory permits us to give a simple interpretation of some of the basic laws describing the behavior of gases. An increase in temperature corresponds to an increase in the mean velocity of the molecules and thus of the energy associated with their motion. Experiments have verified Maxwell's law with great accuracy..."[3]

Of course Clerk Maxwell had come up with this same distribution based on assumptions that Boltzmann was subsequently able to prove were a direct consequence of the conservation laws applied to the collisions of molecules taking place within a gas. Thus, by either of these methods we arrive at a formulation of a probability distribution of velocities of molecules within a monatomic gas.

normalizing the velocity distribution

To determine the values of the integration constants associated with the distribution that was derived in the previous chapter, we posit a constant density ρ of molecules per cubic centimeter throughout the gas as has been shown. Then the velocity density of molecules with a component velocity of v_x in the interval dv_x for example, is simply:

$$\rho_{vx} \, dv_x = A \, \rho \, e^{-v_x^2 / \alpha^2} dv_x$$

---

[3] Prigogine and Stengers, ibid. (p. 241)

And of course the summation of these molecules per cubic centimeter as performed by integration over the entire range of all possible component velocity values independent of the values of the other three component velocities is just ρ, as follows:

$$\rho = \int_{-\infty}^{+\infty} \rho_{vx}\, dv_x = A\,\rho \int_{-\infty}^{+\infty} e^{-v_x^2/\alpha^2}\, dv_x = A\,\rho\,\alpha\,\sqrt{\pi}$$

Therefore,

$$A = 1/(\alpha\,\sqrt{\pi})$$

The probability of a molecule possessing a velocity component of v_x, v_y, or v_z in the respective interval dv_x, dv_y, or dv_z is just, for example:

$$f(v_x)\, dv_x = [\,1/(\alpha\,\sqrt{\pi})\,]\, e^{-v_x^2/\alpha^2}\, dv_x$$

This is, of course, just the 'Normal' probability distribution with $\alpha^2 = 2$.

So the probability of a molecule possessing the speed (magnitude of the velocity) of $v = \sqrt{v_x^2 + v_y^2 + v_z^2}$ in the three-dimensional interval defined as $dv^3 = dv_x\, dv_y\, dv_z$ is just:

$$f(v_x)\,f(v_x)\,f(v_x)\, dv_x\, dv_y\, dv_z = e^{-(v_x^2 + v_y^2 + v_z^2)/\alpha^2}\, dv_x dv_y dv_z /(\alpha^3\, \pi^{3/2})$$

As we saw, $f(v) = f(v_x)\,f(v_x)\,f(v_x)$ and also $dv_x\, dv_y\, dv_z = v^2\, dv\, \sin\theta\, d\theta\, d\phi = dv3$ when converted to spherical coordinates as we've seen, so that:

$$f(v)\, dv^3 = (v^2/(\alpha^3\,\pi^{3/2}))\, e^{-v^2/\alpha^2}\, dv\, \sin\theta\, d\theta\, d\phi$$

Determining the probability that a molecule has a speed (magnitude of velocity) within the interval [v, v+dv] involves multiplication times the volume within the spherical shell lying between the radii v and $v + dv$, which is $4\pi v^2\, dv$. Thus, the probability of a molecule possessing such a speed is:

$$f(v)\, dv = (4v^2/(\alpha^3\,\sqrt{\pi}))\, e\overline{dv}^{v^2/\alpha^2}$$

To assure the distribution is normalized, notice that the integral over all possible speeds must be unity:

$$\int_{0}^{+\infty} (4v^2/(\alpha^3\,\sqrt{\pi}))\, e^{-v^2/\alpha^2}\, dv = 1.0$$

So that:

$$1 = (4/(\alpha^3 \sqrt{\pi})) \int_0^\infty v^2\, e^{-v^2/\alpha^2} dv$$

But since,

$$\partial/\partial\beta\, [e^{-v^2\beta}] = -v^2\, e^{-v^2\beta}$$

therefore:

$$1 = (4/(\alpha^3 \sqrt{\pi}))\, \partial/\partial\beta\, [\int_0^\infty e^{-v^2\beta}\, dv\,]$$

where $\beta = 1/\alpha^2$. The definite integral in the trailing brackets is related to the 'Error function'; its value is just $\frac{1}{2}\sqrt{\pi/\beta}$. Then, performing the differentiation, we obtain:

$$\partial/\partial\beta\, (\frac{1}{2}\sqrt{\pi/\beta}) = -1/4\sqrt{\pi}\,\beta^{-3/2} = 1/4\,\alpha^3\sqrt{\pi}$$

By substitution into the normalization equation, we find that the distribution is indeed normalized.

the normalized velocity distribution features

The resulting distribution is what is commonly referred to as the 'gaussian' distribution. The functional form of this distribution is illustrated in figure 12.1. So let us consider some of the features of this distribution. As a velocity distribution, and we refer here and elsewhere to *velocity magnitude* (speed) unless we specify otherwise, there are several key velocity values whose relationships are illustrated in figure 12.1. These are an *average velocity* designated $<v>$, the *root-mean-squared velocity* defined as $\sqrt{<v^2>}$, and the *most probable speed* which we denote by s_{mp}; they all have unique values as shown.

the most probable speed of molecules

The most likely velocity s_{mp} corresponds to the value where the peak of the functional form of the distribution occurs. That is determined by finding the value for which the slope of the curve equals zero as follows:

$$\frac{\partial}{\partial v} f(v) = (4/(\alpha^3 \sqrt{\pi}))\, \frac{\partial}{\partial v}\, v^2\, e^{-v^2/\alpha^2} = 0, \text{ if } v = s_{mp}$$

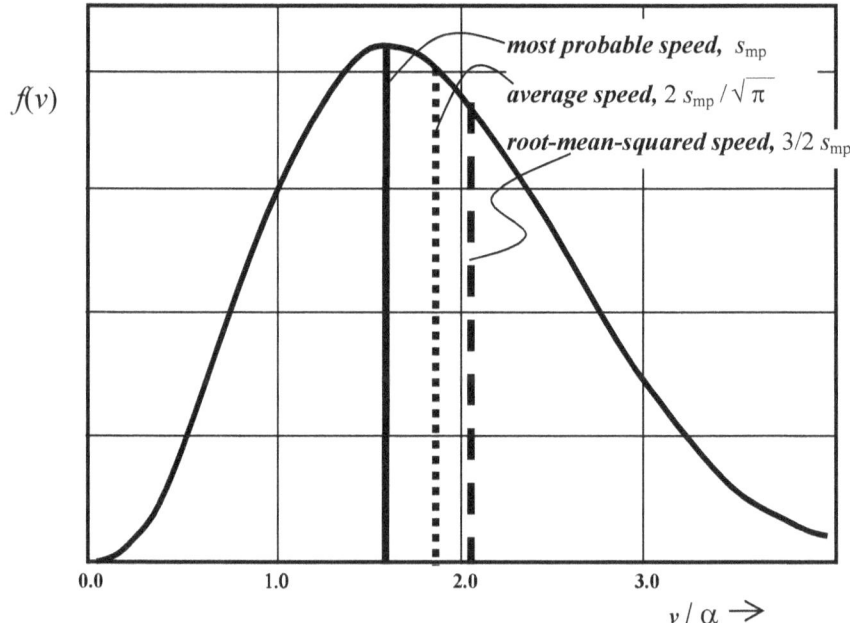

$f(v)$

most probable speed, s_{mp}

average speed, $2\,s_{mp}/\sqrt{\pi}$

root-mean-squared speed, $3/2\,s_{mp}^{2}$

0.0   1.0   2.0   3.0

$v/\alpha \rightarrow$

Figure 12.1: Maxwell-Boltzmann velocity distribution

Performing the differentiation we obtain:

$$(\,2\,S_{mp} - 2\,S_{mp}^{3}/\alpha^{2}\,)\,e^{-v^{2}/\alpha^{2}} = 0$$

Thus we arrive at the value for the most probable speed of:

$$S_{mp} = \alpha$$

the average speed of molecules

Normalized distribution functions in general provide statistical information concerning the parameter for which the distribution function applies, as follows:

$$< g(v) > = \int_{0}^{\infty} g(v)\ f(v)\ dv,$$

where $g(v)$ is a function of v and $f(v)$ dv is the probability of the value of v being in the interval dv.

So trivially the average speed of a molecule in a gas, with $g(v)=v$ in this case, whose distribution $f(v)$ is determined by molecular collisions as derived in the previous chapter is merely the following:

$$< v > = \int_{0}^{\infty} (4 \, v^3 / (\alpha^3 \, \sqrt{\pi})) \, e^{- v^2 / \alpha^2} \, dv$$

$$= (4 / (\alpha^3 \, \sqrt{\pi})) \int_{0}^{\infty} v^3 \, e^{- v^2 / \alpha^2} \, dv$$

Again, by using techniques involving differentiation with regard to the constant in the exponent, we obtain:

$$S_{ave} = \; < v > \; = 2 \, \alpha / \sqrt{\pi} = 2 \, S_{mp} / \sqrt{\pi}$$

the root-mean-squared speed of molecules

Although derived as a velocity distribution, it is more significantly more useful as an energy distribution that applies much more generally than just to the molecular kinetic velocities. So a third type of average is useful because, on the one hand, it is the velocity value that can be determined most directly by experiment using what is known of the molecular masses involved and measured values of the pressure of the thermodynamic system. That is in turn because it is the average energy of the molecules that is reflected in the macroscopic parameters of the system as we demonstrated in chapter 9. The average energy of a monatomic molecule is just:

$$< \varepsilon > \; = \; < \tfrac{1}{2} \, m \, v^2 > \; = \tfrac{1}{2} \, m < v^2 >$$

where $< v^2 >$ is defined as the 'mean-squared' speed.

Thus to determine the root-mean-squared speed we must determine first the average of the square of the speed:

$$< v^2 > = \int_{0}^{\infty} (4 \, v^4 / (\alpha^3 \, \sqrt{\pi})) \, e^{- v^2 / \alpha^2} \, dv$$

$$= (4 / (\alpha^3 \, \sqrt{\pi})) \int_{0}^{\infty} v^4 \, e^{- v^2 / \alpha^2} \, dv$$

Using double differentiation in this case with regard to the constant in the exponent similar to what was done above, we obtain:

$$< v^2 > \; = 3/2 \, \alpha^2$$

And, therefore, the 'root-mean-squared' speed is:

$$S_{rms} = \sqrt{<v^2>} = \sqrt{3/2}\ \alpha$$

the Maxwell-Boltzmann energy distribution

As intimated in earlier chapters, the Maxwell-Boltzmann velocity distribution is interesting primarily because it determines the energy distribution of the constituents of a thermodynamic system. In reference to figure 9.3, we showed that,

$$P\ V = 2/3\ n < \tfrac{1}{2}\ m\ v^2 > = 2/3\ n < E_k >$$

The bracketed expression is the average translational kinetic energy of the molecules striking the wall; n is the total number of molecules. Here P is the pressure and V is the volume of the container. Thus, since the parenthetical expression is just the average kinetic energy of a single molecule, we have that the temperature of a gas is proportional to the total kinetic energy of the molecules. Refer to the ideal gas law formulation for which $PV = NRT$. More specifically, we found that:

$$N\ R\ T = 2/3\ n < E_k >$$

Since N in this equation is the number of moles of the gas and N_A is number of molecules per mole, $n = N\ N_A$. Also, $k = R\ /\ N_A$, so we have:

$$n\ k\ T = 2/3\ n < E_k >$$

Therefore,

$$< E_k > = 3/2\ k\ T = < \tfrac{1}{2}\ m\ v^2 > = \tfrac{1}{2}\ m\ (\ 3/2\ \alpha^2)$$

In the final expression the parenthetical factor is the conclusion of the previous section. From this expression it becomes obvious that α is fully determined by the temperature T of the gas as follows:

$$\alpha^2 = 2\ k\ T\ /\ m,$$

Where $k = 1.38 \times 10^{-16}$ erg/Kelvin is Boltzmann's gas constant and m is the mass of a single molecule.

Now we have the velocity distribution fully determined in terms of physical properties of the thermodynamic system:

$$f(v)\, d\, v = (\, 4\, v^2\, (m\, /\, 2\, k\, T)^{3/2}\, \sqrt{\pi}\,)\,)\, e^{-m\, v^2\, /\, 2\, k\, T}\, dv$$

As derived, it pertains specifically to monatomic molecules. Thus for the Noble gases He, Ne, Ar, and Xe whose atomic weights are 4, 20, 40, and 132 respectively. These values must be multiplied by the atomic mass unit which is 1.66×10^{-24} gm. In figure 12.2 we illustrate plots for the velocity distributions of these gases – all at 298.15 K. This temperature corresponds to 25 °C and 77 °F.

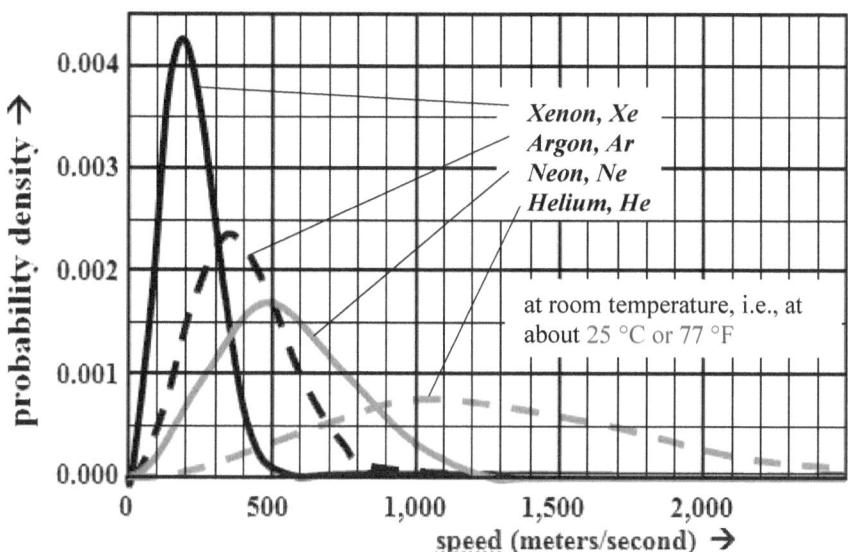

Figure 12.2: Maxwell-Boltzmann velocity distribution for noble gases

The value of the distribution is primarily as a distribution of energy and not just velocities (i. e., translational energy) of the constituents of the system. But it is easy to convert the distribution to energy for the translational situations we have taken into account for monatomic molecules. Using the kinetic energy relation

$$\varepsilon = \frac{1}{2}\, m\, v^2$$

we see that the differential dv can be converted to a kinetic energy differential as follows:

$$d\varepsilon\ =\ m\, dv$$

By substitution into $f(v)\, d\, v$ we obtain the counterpart distribution of kinetic energy:

$$f(\varepsilon)\, d\varepsilon = 2\, (2\, k\, T)^{-3/2}\, \sqrt{\varepsilon/\pi}\ e^{-\varepsilon/kT}\, d\varepsilon$$

contribution of the theory of statistical mechanics

The statistics of the distribution of energy became the later focus of interest. The work of Boltzmann and others in the area of individual interactions in the kinetic theory of gases that had devolved out of the more traditional thermodynamics work was evidently perceived as a dead end with the criticisms of Lochschmidt and others. It became a nonstarter with regard to explaining the nature of entropy and with it, a failure to explain irreversibility arising from what were exclusively reversible processes. There is something about equilibrium and the stable distribution of energy that precludes system states from reverting to states and associated distributions that preceded equilibrium and the kinetic theory could not resolve that issue.

This new focus was to switch the emphasis to 'ensembles' of molecules rather than the individual molecular interactions. It was more or less a natural outgrowth of Boltzmann's efforts. Willard Gibbs and Albert Einstein independently introduced this ensemble theory between the years 1902 and 1904. For his part, Einstein was concerned primarily with the phenomenon of Brownian movement as confirming a molecular level of macroscopic substances, but with both men contributing substantially to the new theory. It would become known as 'statistical mechanics' which is what most physicist now consider the bottom level of analyses in thermodynamics; the kinetic theory seems to be seen as passé.

This theory addressed the dynamics of huge conglomerations of molecules based exclusively on statistics without requiring precise initial conditions. The theory involves a 'phase space' description of an ensemble of particles. Phase space includes the three dimensional locations and three dimensional velocities of an n particle system. So it includes 6 n dimensions. The state of a system can be described by a single point in this 6 n dimensional space; its trajectory through this space describes the successive submicroscopic states of the system. So that as the macroscopic state of a system evolves, the corresponding submicroscopic states transition from point to point. For a single macroscopic state the submicroscopic states move on a trajectory throughout conscribed regions of phase space, all corresponding directly to the one macroscopic state. It is with the characteristics of these regions that the theory concerns itself.

In our earlier discussion we introduced the probability density functions of molecules in 3-space and also of velocities in their own 3-

space. Now there is a similar construct of the density of the vicissitudes of the system's submicroscopic state trajectory visiting points in 6-space over finite time intervals. This density function has superseded what Boltzmann saw as significant in his work. This new density function includes the velocity distribution, but also much more, including probabilities of separation distances of molecules, etc. So it includes virtually everthing there is to know about a system. How could it ever be assessed? It turns out that there are mathematical tools to do this.

Newtonian mechanics evolved to elegant methodologies with which system dynamics could be summarized by a single function called the 'Hamiltonian'. With the Hamiltonian approach the density function can be treated in a similar manner to the dynamics of a compressible fluid. What results is that occupied regions of phase space, however altered, remain essentially the 'same size', i. e., the same number of points involved. And once the system reaches equilibrium, the density function will not change. But the conservation laws still apply to this Hamiltonian approach and the dynamics that are treated are reversible.

afterthoughts and conclusions

It has been shown that the three dimensional component velocities must each separately belong to the same velocity distribution function. Furthermore, as discussed earlier, the pressure produced by molecular impacts with the walls of a container of a thermodynamic system does not depend upon the type of molecules involved. A mole of one gas involves the same number of molecules as that of another independent their mass. We found that by Avogadro's hypothesis that the mean kinetic energy of the molecules of the two gases must be the same although therefore the velocities will not. Dalton's experimental discovery that for a mixture of gases that do not react chemically to each other, the total pressure exerted on the walls of a vessel is equivalent to the sum of the pressure that would be exerted by each if it were to occupy the same volume separately.

Of course these proofs involved the special case of the translational kinetic energy of monatomic gases. However, energy is partitioned much more disparately than that restriction accommodates. The proof that all forms of submicroscopic energy including rotational must all be equal participants among the constituents of thermodynamic systems is most clearly understood in terms of statistical mechanics that we will explore only briefly. Suffice it to say that it has conclusively been shown to be the case. So the use of the parameter ε in the distribution equation above is intended to suggest constituent energy generally.

Einstein was content to show that the Planck distribution of electromagnetic frequencies emitted and absorbed by the molecules within an ensemble is merely *compatible with* Maxwellian distribution of velocities of the same molecules. This is somewhat along the lines of Boltzmann's demonstration of compatibility between collision processes and the energy distribution. Einstein's discovery implied merely that imposing such a blackbody spectrum of radiation of the same internal energy and temperature on a gas that otherwise had no radiant energy for whatever reason, would just maintain the equilibrium characterized by the corresponding Maxwell-Boltzmann distribution of mechanical velocities. But there must be more than mere compatibility; there must be modes of transitioning both back and forth.

That both energy distributions must be compatible certainly makes sense, but just as certainly it puts the kibosh on heat being "nothing more than the motions of the molecules". Einstein did *not* provide the mechanism from which a mechanical energy distribution could *produce* the Planck distribution of radiant energy, nor yet how the Planck distribution might *produce* an associated Maxwellian distribution of particle energies to effect the complete synchronization. Nor did he adequately address the duality problem involving mechanical collisions in turn imposing constraints on the distribution of radiant energies and vice versa. Mechanical and radiant energy were just two happenstance separable energy partitions within the overall internal energy of the substance. They would seem to be on an equal footing counter to what has been implied by initial tentative definitions that associated heat exclusively with 'internal motions' of the particles, i. e., with particle energies, while saying nothing about the associated radiation.

However, since equipartitioning of energy is ubiquitous in substances in equilibrium, the ebb and flow of such a multifaceted partitioning scheme must be an integral part of the process of achieving the equilibrium condition of a substance. That is what constitutes an equilibrium condition.

What are the individual mechanisms that produce the equilibrium between energetic modes? That is the significant issue. What is the thermalization process whereby radiation associated with a thermodynamic system attains its blackbody form? And what are the interaction processes whereby molecules are sped up or slowed down by their interactions with radiation to come into equilibrium with the radiation spectrum?

We must therefore address the hitherto unexplored issue of how a portion of mechanical energy is inevitably converted into thermal radiation. There can be no other primary cause of, and we must insist that it is *not* tantamount to a mere *emergence* of, such thermal radiation. And we must also consider the transfer the other way around. We thus embrace the paradox acknowledged by the underappreciated genius of William James Siddis in the quotation with which we begin our next chapter.

13: Heuristic Models of Entropy

"It would seem, then, as though there must be some reason in terms of the reversible physical laws why the second law of thermodynamics must be true; that is, the second law of thermodynamics, if true, should be a consequence of the reversible physical laws applicable to ultimate particles. We are, then, confronted with the paradox of having to deduce an irreversible law from perfectly reversible ones."[5]

"Boltzmann, however, wanted to go farther. He wanted to describe not only the <u>state</u> of equilibrium but also <u>evolution</u> toward the Maxwellian distribution. He wanted to discover the molecular mechanism that corresponds to the increase of entropy, the mechanism that drives a system from an arbitrary distribution of velocities toward equilibrium."

"...Can we conclude that the problem of irreversibility has been solved, that Boltzmann's theory has reduced entropy to dynamics? No, it has not.."[9]

It is clear that Boltzmann devised this 'H theorem' to prove the irreversible succession of submicroscopic states to a stable equilibrium. It was the culmination of his effort to achieve a reductionist explanation of entropy. The theorem is not even required in a milder form of Boltzmann's derivation of the Maxwell-Boltzmann distribution.

---

[5] William J. Siddis, The Paradox," ***The Animate and the Inanimate***, available at http://www.sidis.net/ANIM4.

[9] I. Prigogine and I. Stengers, ibid. (p. 241-243)

Realization of the stable Maxwell-Boltzmann distribution is merely *sufficient* but not a *necessary* condition for equilibrium; various anomalous conditions can also result in stability as demonstrated by Prigogine and Stengers. But irreversibility makes no entry whatsoever in the reduction of thermodynamics to the kinetic theory by Boltzmann's analyses. Irreversibility had slipped through his fingers.

What the *H* Theorem does imply is that, although there might possibly be some *other* (even if previously unknown) distribution that pertains in some situations of equilibrium, certainly the realization of *Maxwell's distribution does indeed imply equilibrium* in any case. This statement is easily proven by Boltzmann's analysis but the reverse is not. So does the H theorem imply that there is indeed a necessary trend of microstates into only those for which the exponential distribution function applies? No.

The less stringent, but still extremely powerful, constraint is worth some thought. Furthermore, there are well known phenomena such as chemical clocks for which the more stringent *necessity* condition obviously does not apply. In such cases a form of reversibility would seem to be realized to within the accuracy of Boltzmann's analyses even at the macroscopic level of our everyday existence. The *sufficiency* condition simply implies that whether velocities are reversed, which in most conceptions is tantamount to time reversibility, or not, the system will indeed proceed to, or remain in, a 'stationary state' characterized by conditions that typically do, but need not in all cases imply the Maxwell-Boltzmann distribution.

Another aspect of thermodynamic irreversibility is that involving the gradual dissipation of energy from (or to) a local system into (or out of) a larger global system or the entire universe. Here the local system may remain always in an instantaneous quasi equilibrium, but will nonetheless suffer irreversible losses of energy over time until and unless it is in a totally isolated static equilibrium with its global environment. Certainly these two aspects of irreversibility may produce similar phenomena and involve much commonality in explanation, but they pertain to quite differently described physical situations.

the association of 'H' with entropy

We noted earlier Boltzmann's definition of the parameter *H* as a part of what he called his *H Theorem* in his rigorous derivation of the Maxwell-Boltzmann distribution of molecular velocities based on the conservation laws applicable to molecular collisions. As we saw, H is defined as:

$$H = \int_0^\infty f(v) \ \ln(f(v)) \, dv$$

Here v is the magnitude of the velocity (speed) of molecules within a gas and $f(v)$ is the distribution of those speeds throughout the gas. As discussed previously, he showed that,

$dH/dt < 0$.

This conclusion is reminiscent of claims for entropy S except that entropy was conventionally defined in a negative sense such that,

$dS/dt \geq 0$.

So it was quite natural to wonder whether H might be a submicroscopic classical mechanical descriptor of entropy itself. In which case, a reduction of the macroscopic state property entropy would have been achieved as anticipated earlier in determining effects of rapid change in pressure on the energetics of the molecular constituents of a gas.

It can be shown by rigorous analyses based on Boltzmann's work, but getting into the discipline of statistical mechanics which is beyond the scope of the current investigation, that for a monatomic gas in equilibrium, the following relationship does indeed hold:

$S = -kH$ *Entropy*

We will presently use heuristic arguments to provide some measure of confidence in this conclusion.

Thus, it is possible to reduce the *description* of entropy to the submicroscopic level by a completely mechanically-based expression for entropy. Although Boltzmann's achievements were indeed significant, including much of what we currently know about thermodynamics – heat and entropy in particular – he could not adequately explain the inevitability of increases in entropy in approaching equilibrium since the reversible conservation laws from which equilibrium derives do not preclude reversal of that progress. So although on the one hand, his association of H with entropy was persuasive, nonetheless his H theorem could not be applied to convincingly prove the inevitability that is the hallmark of entropy.

the criticisms of Boltzmann's conclusion

Boltzmann later had to conclude, that even this proof could not account for irreversible changes in mechanical systems even though his theorem had captured the nature of entropy as a description even though certainly not as an explanation of its cause or origin.

If one were to reverse velocities of all particles (what had been considered tantamount to reversibility) at any point in time in accordance with Lochschmidt's criticism of Boltzmann's precipitous claim of explanation, the theorem would still apply. Entropy would as surely increase into the past and not uniquely into the future as the arrow of time indicates that it must. Increasing entropy would not have been exclusively associated with irreversible behavior. Thus, the long sought explanation was ultimately to elude Boltzmann. His effort certainly had resulted in a brilliant derivation of Maxwell's equilibrium velocity distribution of the constituent molecules. But the result totally obscured any rationale with regard to distinguishing the issues of reversibility and irreversibility.

The plausibility of the precise balance he had seemed to prove must be considered. One cannot, of course, imagine a counter process of the particular elastic collisions he analyzed whereby an equilibrium so achieved would be frittered away, although that would seem to be a reasonable reticence from Lochschmidt's critique. There is something that Boltzmann and his critics alike had overlooked. (We will find that later in analyses of mediated interactions and scattering phenomena.) Both the analysis of the dynamical processes and the critique seem intuitively valid although there was at the time no discernible flaw in either. The flaw of course is the omission of the significant role radiation plays in the mediation of the exchanges between molecules that is outside the scope of the model implemented by Boltzmann.

So although on the one hand Boltzmann did discover a tremendous amount of useful information concerning behavior of submicroscopic constituents of gases in equilibrium, he most certainly did not discover the origin of irreversibility with which he had become obsessed.

the significance of Boltzmann's work

Let us consider Boltzmann's impression of the significance of his 'H Theorem' in this regard. He realized that his model involved completely reversible processes and yet felt convinced that he could demonstrate how these completely reversible mechanical operations were ultimately responsible for the origin of the notably irreversible behavior of the second law of thermodynamics.

Certainly it was not implicit in Boltzmann's understanding of phenomena at this level of analysis that a molecule becomes 'hot' by the impact of another molecule to thereby radiate thermal energy that would not otherwise be emitted as a merely unrelated normal atomic emission. But something similar to that must have seemed to him like what must happen in actuality. The concept of heat must always have been entwined with the colors given off in thermal radiation and yet that aspect of what heat 'is' never seemed to be exploited. There are clearly other aspects of heat than just molecules in motion. We will address this specifically farther on.

Boltzmann believed that he had found the source of irreversibility in his 'necessary and sufficient' condition. His efforts with regard to the concept of irreversibility exhibit a couple of most interesting facets. One is that associated with Boltzmann's 'H theorem' whereby a system of particles that continues to satisfy the reversible conservation laws tends irreversibly toward an equilibrium characterized by the Maxwell-Boltzmann distribution of energies. It can't. Since persistence of the conservation laws was assumed throughout Boltzmann's derivation, there was assumed to be no dissipation by radiation of total energy. So the extreme initial optimism is hard to comprehend.

He defined the quantity H as the integral (or the summation in discrete finite element systems) of the composite distribution function $f(v)$ times the logarithm of $f(v)$ where integration (or summation) is over all possible velocities, v. In deriving the function $f(v)$ representing the distribution of energy apportioned to individual particles of the system, we saw that it must be an exponential function when the system is in equilibrium:

$$f(v) = (2 \pi kT)^{-3/2} e^{-\frac{1}{2} m v^2 / kT}$$

$$\ln f(v) = -\frac{1}{2} m v^2 / kT - 3/2 \ln (2 \pi kT)$$

So naturally, the integral over all possible velocities of the Maxwell-Boltzmann distribution times the log of that distribution is always negative except at absolute zero where the molecular velocities would be zero. This is a proof of Boltzmann's H Theorem in a very particular instance, not a proof that the H theorem has any particular significance.

combinatorial aspects of energy distributions

We proceed now to considerations of the combinatorial problem involving the number of equivalent combinations of energy states at the submicroscopic level of molecules in a system with given energies in an unchanging distribution. It can be shown that this number of equivalent combinations increases as a system tends toward equilibrium even though the total energy of the system remains the same.

It is largely the relative velocity v that constitutes the energy of a molecule in a gas that will be of concern to us here, since classically the kinetic energy of a particle is $E_k = \frac{1}{2} m \mathbf{v}^2$. The atomic and molecular particles in a homogeneous gas are all the same, so their rest masses are all equal. Thus, since it is the square of the velocity that appears in the expression above and in relativity, it is the 'speed' of a molecule, $v = |\mathbf{v}|$, independent of its direction that affects its energy. To the extent that the submicroscopic particles are indivisible, i. e., the system is not at or near a temperature for which particles experience chemical bonding, disassociation reactions, or any thermonuclear forces, it is primarily this kinetic energy expression that applies. For the more complex cases, that we will not directly address here, the additional rotational energy components would need to be taken into account in the same way as well. But even if we ignore those complications, a completely even distribution of energy implying (in our limited case) an even distribution of particle speeds would be one for which every particle experienced the very same speed relative to its confines, which is so unlikely for any appreciable number of particles as to be totally inconceivable. And if it were at one particular moment *possible*, after a single collision it could no longer be the case.

So what is meant instead when we speak of a 'uniform distribution' is one for which velocities are 'stably' distributed, i. e., if a high speed particle is slowed down by a collision, there must be a slow speed particle somewhere in the system that is sped up by colliding with another particle to take its place in the overall distribution. The number of particles with any given energy will not change over time if only elastic collisions are involved. Boltzmann demonstrated that this kind of stability results for the Maxwell/Boltzmann distribution. Clearly, the conservation laws prove it to be so, but they do not drive it there..

In any case, to understand entropy and irreversibility one must understand the mathematical aspects of how one assesses the likelihood of a molecule possessing a given amount of the total energy of the system. It is that distribution of the total energy among submicroscopic particles that remains stable even as the particles themselves collide and exchange the energy they possess.

To comprehend the extreme complexity involved in the distribution of energy, consider the amount of diversity in possible 'placement' of particles into categories associated with the amount of energy assigned to a given particle. A particle with no energy could be allocated to a category labeled '0', for example. A particle with one unit of energy would then be allocated to a category labeled '1', a particle with an energy of two units would be allocated to a category labeled '2', etc.. Notice that we are initially addressing only discrete units of energy 0, 1, 2, etc. times some discrete value ΔE. Later we will address extending this analysis to a continuous spectrum of energy categories.

Suppose we have a system with total energy of all included particles of 10 ΔE. In distributing the particles among allowed energy levels, the most uniform distributions of energy among the particles would be the ones in which most of the particles share some of the energy. Since we assume the particles are identical, the order of the particles assigned to the same energy level is irrelevant – one ordering is indistinguishable from all the others. We are interested in exploring the ramifications of increasing the number of particles to share the total energy:

If there were but one particle, it would have to be allocated ten units of energy and placed in category '10'; there is only one way to do that. If there were two particles, we could still allocate all the energy to one particle, there are two ways to do this, one particle with 10 units and whichever other one with zero. Or we could allocate 9 units to one particle and 1 unit to the other with only two ways to do this. We could also choose one particle with 8 units and one with 2 units, or one with 7 and the other with 3 units, or an allocation of 6 and 4 with only two ways to do any of these. On the other hand we might allocate 5 units to both particles; there is only one way to do this. Thus, there is a total of 11 ways to distribute the ten total units of energy between two particles.

With three particles, we could again allocate all the energy to one particle with two particles allocated 0 units; there are just 3 ways in which to do this. Or we could allocate 9 units to one particle and 1 unit to another with zero allocated to the third; there are $6 = 3 \cdot 2 \cdot 1$ ways to do this. We could also choose one particle with 8 units, one with 2 units, the third with 0. Or we could allocate one with 7, another with 3 units, with 0 allocated to the third, or a permutation of 6, 4, 0, or we could allocate 5, 5, 0, for each of these possibilities there are 6 possibilities. This gives a total of 30 permutations with two particles sharing all the energy. (Notice that the 4, 6, 0 allocation is included in the 6, 4, 0 accounting, etc..) We could also do the allocations 8,1,1; 7,2,1; 6,3,1; 6,2,2; 5,4,1; 5,3,2; 4,4,2; or 4,3,3. There are also 6 ways to achieve each of these allocations, for a total of 48 permutations with all three particles sharing some portion of the total energy. Thus, for three particles there are a total of 81 ways to allocate the ten units of energy.

We see that with increasing numbers of particles in the system the more that even distributions of the total energy predominate. Thus, if there were 55 units of energy in the system and ten total particles there would be ten distinct configuration possibilities for which one particle could be allocated all 55 units. The most uniform allocation would be one in which each energy category has exactly one particle. In this case there would be well over 3 million possible arrangements all involving exactly the same total energy. To elaborate for this case we have ten choices for the placement of the first particle, i. e., particle number one could have an assigned energy of 0, 1, 2, up through 10 units. Since we are considering an even distribution, the next particle must be placed in a different category than the first, so there are only nine choices for the placement of the second particle, eight for the third, and so on. So there are a total of *ten factorial* choices – symbolically, 10!

$$10! \equiv 10 \cdot 9 \cdot 8 \cdot 7 \cdot 6 \cdot 5 \cdot 4 \cdot 3 \cdot 2 \cdot 1 = 3,628,800$$

This is only the most uniform distribution of 55 units of energy among ten particles. Of course we could allocate twenty seven units to each of two of the particles with one particle having one unit and the others no energy at all, etc.. The number of possibilities of each of these is quite enormous, but none is as large as for the most uniform. In every case, the more uniform distribution has a larger number of possible realizations.

This illustrates that there are literally millions of ways to obtain an equivalent uniform energy distribution of as few as ten particles. So for real systems in which the number of particles and possible units of total energy is astronomical, the number of ways for which an equivalent distribution of energy can be obtained becomes truly a mind-boggling uncountable number. Grappling with that extreme number of possibilities is the starting point for defining entropy at the submicroscopic level.

To simplify the discussion, suppose we have a system of just 5 particles with a total of 5 possible discrete units of energy to be shared among the 5 particles. Notice that this does not allow inclusion of all combinations of the five particles with the same number of units of energy because most of those combinations do not comply with the total energy constraint on the proposed system. For example, the assignment of five particles with one unit of energy each is viable whereas any other combination of five particles with the same energy is not. So the first question is, "How many unique combinations of this limited resource of five units of energy shared among five particles are there?" As stated above, there is only one way for all five particles to possess the same amount (one unit) of energy and still meet the constraint of a total of 5 total units of energy. In Table I we enumerate every one of the viable combinations, since in this very limited case of five units of energy shared among 5 particles it is a feasible task to accomplish.

TABLE I

combination description (number of particles sharing given number of energy units)	number of combinations # ways	number of particles per energy unit partition #					
		0	1	2	3	4	5
1 particle with 5, 4 particles with 0	5	20	0	0	0	0	5
1 particle with 4, 1 with 1, and 3 with 0	20	60	20	0	0	20	0
1 particle with 3, 1 with 2, and 3 with 0	20	60	0	20	20	0	0
2 particles with 2, 1 with 1, and 2 with 0	30	60	30	60	0	0	0
1 particle with 2, 3 with 1, and 1 with 0	20	20	60	20	0	0	0
5 particles with 1	1	0	5	0	0	0	0
TOTALS	**96**	**220**	**115**	**100**	**20**	**20**	**5**

It is a premise of these calculations that any possible permutation of matching specific numbers of particles with specific amounts of energy are all equally likely as long as the total energy of all particles in the system is conserved. So it is imperative that we determine the number of possible permutations of allocating energy to particles for each primary combination in order to figure out the probability of particles possessing a specific amount of energy.

To understand the determination of how many possible permutations involve particles being allocated a given number of units of energy in these combinations let us consider the case associated with row number two in Table I. In this combination a single particle possesses 4 of the units of the total of 5 energy units, another one possesses 1 unit, and the other 3 of the particles possess no units of kinetic energy at all. This is true in every one of the combinations of which there are $20 = 5 \cdot 4$ total. Notice that in these 20 permutations there are $60 = 5 \cdot 4 \cdot 3$ ways in which a particle could have zero units of energy.)

Why there are 20 unique permutations (and exactly what those permutations are) in row number two is obvious as illustrated in Table II below. Tables similar to Table II could be constructed for each row in Table I to verify the numbers provided in the six far right columns. Thus, we arrive at there being 220 equally likely ways for one of the five particles to be allocated zero units of energy, 115 ways for a particle being allocated a single unit of energy, etc.. Clearly, there are many more ways for particles to possess small portions of the total amount of energy than there are for it to possess larger portions of the total amount.

TABLE II

The 20 permutations of the combination of one particle with
four units of energy and one with a single unit of energy

particles allocated to each number of units of energy						
permutation	0	1	2	3	4	5
#1	3,4,5	2			1	
#2	2,4,5	3			1	
#3	2,3,5	4			1	
#4	2,3,4	5			1	
#5	3,4,5	1			2	
#6	1,4,5	3			2	
#7	1,3,5	4			2	
#8	1,3,4	5			2	
#9	2,4,5	1			3	
#10	1,4,5	2			3	
#11	1,2,5	4			3	
#12	1,2,4	5			3	
#13	2,3,5	1			4	
#14	1,3,5	2			4	
#15	1,2,5	3			4	
#16	1,2,3	5			4	
#17	2,3,4	1			5	
#18	1,3,4	2			5	
#19	1,2,4	3			5	
#20	1,2,3	4			5	
totals	60	20	0	0	20	0

But exactly how does this relate to entropy and then ultimately to irreversibility? We have yet to explain that.

We might simply have attempted to define the entropy of a given distribution of energy as proportional to the number of separate ways that a distribution could be comprised. Refer back to column 2 of Table I, where the number 96 would be the defined value of entropy in that case. However, if that had been done, the total entropy of two similar isolated systems that are then to be considered as a single system would have to be obtained as the product of their separately determined entropies. This is because the two systems would have to be regarded as distinct and therefore, since each of the possible arrangements of particle energies of the first system's distribution would have to be combined with each of the possible arrangements of the second system in its distribution, the total entropy of the two would then be the product of their respective entropies.

Definitions must be made in such a way that they have practical application to the problems at hand. And, it is more useful to define entropy such that the total entropy of two such systems is merely the sum (rather than the product) of their individual entropies. However, it is essential to retain the basic rationale that we associate with combinatorial probabilities as discussed above. With an operative definition that takes both these advantages into account, entropy could then be considered an intrinsic property of a thermodynamic system as has been assumed and discussed as such in the earlier chapters of this book.

To retain these desired aspects, consider the fact that the logarithm of a product is equal to the sum of the logarithms of the individual factors. Thus, the advantages sought can be realized by defining the entropy S of a distribution with n equivalent configuration possibilities as:

$$S = K \ln(n),$$

where K is a constant and $\ln(n)$ is the natural logarithm of n. Then the entropies of the two system distributions considered as a single dual system just discussed would be defined as:

$$S_1 = K \ln(n_1) \text{ and } S_2 = K \ln(n_2),$$

And thus, the entropy of the combined distributions would be:

$$S_{12} = K \ln(n_1 \cdot n_2)$$

This becomes:

$$S_{12} = K \ln(n_1) + K \ln(n_2) = S_1 + S_2$$

We have thus retained the feature of the product as the number of permutations of particle energies in the overall combined system in our definition of entropy which has the desired extensibility feature.

From the previous discussion, a permutation in which all ten particle energies were of unit value is in some sense equivalent to one in which all ten particles shared a different energy value, even though we recognize that they correspond to ten very unique systems. Ten particles, all with energies between 9 and 10 ΔE, is most definitely not equivalent to ten particles, all with speeds between 0 and 1 ΔE from a thermodynamic perspective. Thus, in counting numbers of configurations, each particle as well as each energy level must be considered as a distinguishable entity. Thus particle energy levels *cannot* be considered interchangeable.

associating 'micro' and 'macro' system states

Clearly there are considerable conceptual difficulties in assigning a total number of unique configurations of particle energies that might comprise a given total energy distribution. This is all part and parcel of

143

making the reduction from the macro to the micro level descriptions of a system. Individually unique arrangements of particle energy are what we have referred to as the microstates of a system whereas the overall distribution of energy is a somewhat higher level description of the state of the system. Thus, it makes sense to define the entropy of a given macrostate as proportional to the logarithm of the total number of microstates that might equivalently comprise that system's state. It is assumed here that each microstate (a single row in Table II for example) is equivalent in the only meaningful sense of contributing exactly one instance to the total number of microstates compatible with a single system state. We don't uniquely qualify microstates in this regard; we just count them.

This concept of entropy was developed in relation to ideal gases. So in the context of the particles discussed above, there would be a fixed number, n of particles which are gas molecules in the 'system' but with a rather extensive range of possible velocities (energies). A 6 n-dimensional 'phase space' of microstates can be defined for any such system, where for each of the n molecules there is a 3-dimensional position and a 3-dimensional momentum (velocity times mass of the molecule) 6 n dimensions in all. In statistical mechanics the concept of a unique microstate of the system is represented by a single point in this 6 n-dimensional 'phase space'. In defining the entropy of a macrostate of a system, we partition the phase space of microstates into regions of 'equivalent' microstates, any one of which might comprise the state of the system. Then the number of microstates contained within this region is defined as the entropy of the system in that state.

The distinction between 'macro' and 'micro' is somewhat subjective in the first place as we have seen, so of course defining the boundaries of the numerous microstates representing single macrostates will be somewhat arbitrary. But we must stand by this meaningful many-to-one distinction because otherwise the entropy of every state would be zero, since $\ln(1) = 0$. The practicality of the concept of entropy as addressing the problems of irreversibility depends upon there being such an aggregate property that takes into account all of the lower level possibilities of each macroscopic higher level state of a gas as defined by temperature, volume, and pressure. Aggregate regions of phase space that correspond to highly uniform distributions of energy are extremely large since they contain many more permutations of particle positions and velocities of equivalent energy. The number of permutations associated with a stable equilibrium distribution of any nontrivial quantity of a gas is many orders of magnitude larger (i.e., they have very many more points in phase space) than do the regions corresponding to non-equilibrium distributions.

Typically a single macrostate of a system vacillates through the phase space of a vast number of microstates with particles exchanging their energy, momentum, and position, as the system progresses sequentially from one point in this space to a next, etc., ultimately defining what one would have to think is an ever larger and larger region of the space that has been occupied by the system at one time or another. The conservation of energy guarantees that these all involve the same amount of energy.

Clearly, there must be some directionality toward a somewhat limited *safe* place to this *meandering* through phase space to imply irreversibility as Boltzmann came to assume. And there must be no return from this region once it is attained. But how?

resulting combinatorial distribution of energy

So far we have discussed the combinatorics of allowed permutations of energy assignments to particles. But let us look now at the distribution of energy among the particles independent of individual permutation more or less as Boltzmann had. In the final five columns of Table I we have tabulated the numbers of instances of particles being assigned a given number of units of the total energy. Since each viable permutation of energy assignment that satisfies the total energy constraint must be considered equally likely, we can change these numbers to 'frequencies' (i. e., probabilities), to obtain:

$p(0) = 220 / 500 = 0.44$

$p(1) = 115 / 500 = 0.23$

$p(2) = 100 / 500 = 0.20$

$p(3) = 20 / 500 = 0.04$

$p(4) = 20 / 500 = 0.04$

$p(5) = 5 / 500 = 0.01$

Here $p(x)$ is the probability of a particle having the energy x. If we plot these numbers for the five-particle five-unit-of-energy system, we obtain the darkened bar chart shown in figure 13.1 below.

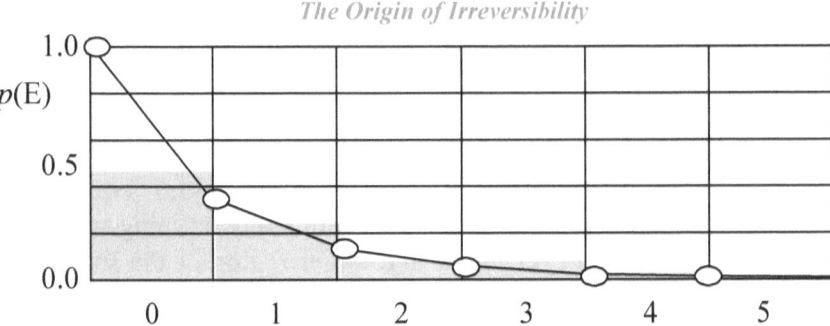

Figure 13.1: Probability of particle possessing a given energy

The open ovals represent the value of the exponential function that such bar charts will increasingly approximate as the number of particles and the number of total energy units becomes extremely large with energy unit size decreasing accordingly. With such an extension, associated bar charts will approach a continuous function of probability of the given energy per particle as a function of given units of energy. This function is the probability of a particle possessing that particular energy. The continuous function can be expressed as:

$$p(E) = A\ e^{-\sigma E}$$

where A and σ are constants of the system. The value of $\sigma = 1\ /\ <E>$, where $<E>$ is the average value of the energy of all the particles.

$$<E> = E_{TOTAL}\ /\ N$$

Here N is the total number of particles in the system. The value of A is determined by requiring that the sum (integral) of the probabilities over all possible energies is equal to unity as required by any normalized probability distribution function, so that:

$$A = 1\ /\ \Sigma\ p(E_i) = 1\ /\ \int e^{-\sigma E}\ dE$$

So in our case,

$$A = 1\ /\ \sigma = <E>$$

This distribution function is the same as that derived by Boltzmann based on considerations of the conservation of energy and momentum that must persist throughout particle collision impacts. We have already demonstrated his derivation in chapters 11 and 12.

Suppose for now that we were to infuse a jet of high speed molecules into a container of gas and then seal it off again. In so doing we would increase the total number of particles and the total energy contained in the gas as well as introducing a very non-uniform distribution of particle velocities/energies. The initial temperature, pressure, and volume of the gas would be indeterminately altered through a succession of macrostates compatible with the newly increased internal energy as the infused particles gradually transfer their higher energies to other molecules in the gas. This would produce a sequence of different distributions of particle velocities associated with different macrostates as well as different microstate permutations. Eventually this rapidly changing sequence of microstates should transform the system to a new stability. Ultimately this should result in a uniform stable Maxwell-Boltzmann distribution of molecule energies associated with a new macrostate with altered temperature, pressure, and volume.

But is that necessarily true? As was illustrated in figure 11.2, collisions do not redistribute relative velocity; after each impact the same relative velocity difference between molecules pertains, so how is it that the overall distribution would change?. Also, having once reestablished a new stability, we expect the system to remain in that newly realized equilibrium state unless and until it was forcibly altered once again. We intuitively expect that thereafter as the particles interact resulting microstates will all be compatible with the newly established macrostate without further changes. No new macrostate should arise once the new stability has finally been established. But if reversible interactions were what produced the new stable state, how will they not fritter it away?

We will later show how photon mediation of molecular interactions makes subtle but significant changes that do, in fact, produce the ineluctable trend to a stable state of equilibrium.

How do number of permutations imply irreversibility?

Since the particle energy and momentum exchanges that precipitate new permutations according to Boltzmann's model are envisioned as resulting directly from the reversible laws of mechanics, we encounter a situation whereby the same exchange considerations must apply even if time were to be reversed. In other words, if we reverse each interchange between every particle starting at the present, we could follow the interaction sequences backward in time to where the infusion discussed in the previous paragraphs had occurred. Ultimately the injected gas would be ejected. That's what reversible laws imply. The very same Newtonian laws of physics are operative in going both ways at the submicroscopic level, and should therefore produce ebb and flow of entropy rather than resulting in a unilateral decrease into the past and increase into the future.

Thus, if we maintain that the entropy of every system, including the entire universe, has been continuously increasing, leading up to the present moment, we are forced to conclude that the sequence of interactions that are associated with a system 'meandering' through phase space must not be reversible for some unknown reason. The phase space boundaries surrounding microstate regions associated with stable distributions of the overall system must somehow be inviolable unlike the boundaries that were trespassed in getting from an infusion of energy to a new stable state. The only other alternative is that the laws of physics operative at the microscopic level are, in fact, not temporally symmetric. Again we face the question of how a system whose behavior at the submicroscopic level has been unabashedly assumed to be governed by symmetrical laws must proceed asymmetrically in this way. The facts refute the conjecture.

Many proposals have been made to resolve this enigma. They are, as we have seen, mostly of two general types. One is the argument that the second law of thermodynamics is 'explained' by the mere (assumed) 'fact' of extremely low entropy in the past. Alternatively, it is argued by some (including the authors) that the fundamental laws of physics as they play out in the interactions of basic particles are *not* in actual fact temporally symmetric, despite their apparent reversibility. Both 'explanations' address the problem, but neither has to date provided a plausible mechanism to justify either claim.

The entry of irreversibility was argued by Walter Ritz to derive from a fundamental flaw in Maxwell's equations of electrodynamics because they permit 'advanced' as well as the usual 'retarded' wave solutions. He argued that since we never observe these advanced waves moving into the past, there must be something wrong with the equations that permit such solutions, which thereby must be filtered to exclude the nonphysical solutions. Proceeding from classical to modern wave theory we note that Schrodinger's wave equation is also reversible in a similar sense, although the separable measurement process in quantum mechanics could be considered to be an irreversible process. But none of these arguments provide a meaningful explanation for how irreversibility comes about naturally by any such mechanism, i. e., how it plays out in particle interactions rather than being artificially imposed as a separate scientific discipline. That determination is the holy grail that no one has so far found.

Whatever the case, it clearly is not possible to propel the state of a system backwards in time symmetrically with processes that are otherwise-similar to those that proceed forward in time. This is not just because this leads unavoidably to the implication of greater entropy in the past, which it would, but because it implies phenomena that have never been observed. It must be that either there is some hitherto unacknowledged temporal asymmetry in the physical laws, there is some physical constraint on microstates in phase space that prohibits them from transitioning to states associated with less uniform macrostate distributions which is tantamount to the same thing, and/or there is a unidirectional uncertainty in the microstates of particles that precludes precisely reversing their interactions to progressively obtain lower and lower entropy.

On scales of a few atoms, of course the second law, and indeed thermodynamics itself, does not strictly seem to apply. Whether this is just because we are confident that we understand everything that is going on in such systems and it does not jibe with what we see as thermodynamics, or whether for some reason thermodynamics just *doesn't* apply, is perhaps worth thinking about. In any case, as suggested by the introductory quotation from Weinberg to start chapter 1, such tiny systems are for whatever reason considered outside the domain of thermodynamics proper. But... they are of extreme significance to the study of the kinetic theory of matter, thermal energy, and statistical mechanics, and therefore to thermodynamics itself. Typically such small-scale systems are investigated from the distance of statistics and probability as in the energy distributions of such interactions in the kinetic theory of gases or in even more abstract statistical mechanics. And when subjected to such summarizing methods, significant individual microscopic particle interaction phenomena must surely be lost.

What is the solution?

Relativistic and photoelectric effects discovered after much of the kinetic theory research was performed alter the picture of submicroscopic behavior. Electromagnetic radiation is subject to Doppler effects that mediate particle collision processes. Plus, there is transverse component of uniform relative velocity that unilaterally reduces expected energy exchanges between molecules in relative motion. Although conservation laws certainly still apply, something is lost in the exchange. The relative velocity is reduced by the exchange. The incommensurability of energy and momentum relationships that apply to particles and those that apply to radiation is very significant as we will see. These obvious but little appreciated aspects of molecular interaction mediation will ultimately be identified as culpable processes in the discussions of later chapters of this volume.

The detailed evidence for this conjecture will be treated exhaustively and its conclusion verified. This discovery warrants a renewed optimism in the reductionist agenda. It provides a fuller understanding of heat, entropy, and various other concepts associated with thermodynamics. And it augurs well for other scientific endeavors where *emergence* or other pseudo-scientific hocus pocus seems increasingly to have usurped authority. It provides hope for such areas previously heralded as at an impasse with regard to legitimate reductionist explanations.

Part III

14: 'Modern' Physics Amendments

"...Speaking on December 14, 1900, at the meeting of the German Physical Society, Max Planck stated that paradoxes pestering the classical theory of emission and absorption of light by material bodies could be removed if one assumed that radiant energy can exist only in the form of discreet packages. Planck called these packages light quanta. Five years later, Albert Einstein successfully applied the idea of light quanta to explain the empirical laws of the photoelectric effect....."[3]

So how does one address the various types of interaction between particles and radiation that were not considered by Boltzmann? This is particularly problematic because relationships between the expressions for energy and momentum of particles and those for radiation that are essential to theoretical models of 'modern physics' differ considerably. The associated conservation equations exhibit features that seem to be at odds with one another and limiting the interactions that are viable. This certainly complicates attempts to integrate models of separately-considered particle and radiation behavior.

How do molecules and radiation interact?

In classical physics electromagnetic radiation was acknowledges as possessing both energy and momentum. But that acknowledgement did not include anything like what has become familiar to virtually everyone since early in the 20th century with regard to what light (or electromagnetic radiation generally) really *is*. Refer to figure 14.1.

---

[3] G. Gamow, *Thirty Years That Shook Physics*, Dover, New York, 1966, p. 1.

It was known that atoms (and therefore molecules made up of them) were comprised of positively charged nuclei (with neutral neutrons and/or positively charged protons) as well as the negatively charged electrons. When an electron was 'attracted' to a nucleus, it was supposed that it spiraled in with radiation rippling out more or less like the ripples in water emanating from the spot at which a rock had splashed. So the electromagnetic energy of the separated positive and negative charges was envisioned as though continuously emanating in all directions. The energy and momentum associated with the radiation was emitted in all directions and the symmetry significantly simplified the solution of conservation of energy and momentum equations.

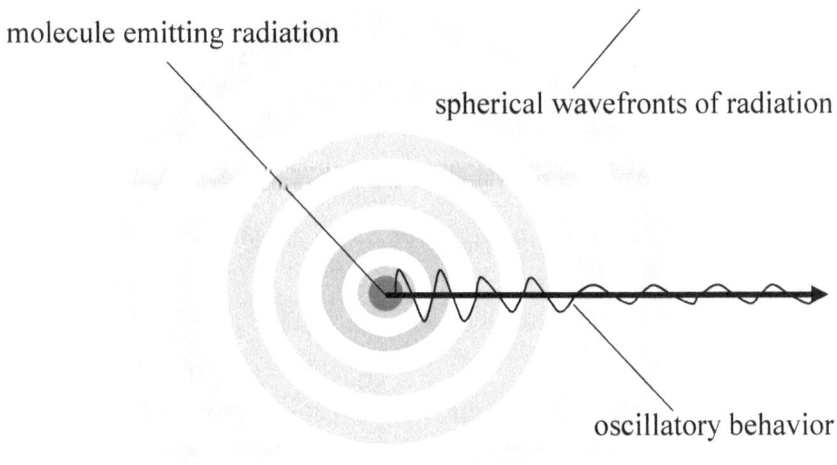

molecule emitting radiation

spherical wavefronts of radiation

oscillatory behavior

Figure 14.1: Earlier classical view of spherical wave function radiation

What is a photon?

Among Albert Einstein's major discoveries during the first decade of the 20th century was the 'photoelectric effect'. The significance of this discovery was that it demonstrated that light transmission involves discreet chunks (Plank's 'quanta') of energy, 'photons' as G. N. Lewis would later denominate them. The further significance is that matter can only absorb entire photons of electromagnetic energy. We will see later in discussing Einstein's quantum theory of radiation, that a precise balance must be maintained, requiring precisely limited amounts of energy to be absorbed in transitioning the molecule to a next energy level with no 'left over' radiation being left without absorption.

However, we must acknowledge that the specific energy and momentum of the photons that effect the transitions between energy levels in matter are not quite as pristine as we have just implied. The *uncertainty principle* of quantum mechanics loosens these restraints. This principle states in particular that there is a lower limit to the product of the uncertainty in the energy ΔE times the uncertainty in the transition time duration Δt of the energy that is being transferred. This limitation cannot be reduced to less than a constant times Planck's constant h as follows:

$$\Delta E \, \Delta t \geq \frac{1}{2} \hbar,$$

Here the symbol $\hbar$ is pronounced 'h-bar' and is defined as:

$$\hbar = h / 2 \, \pi.$$

where h is Planck's constant. Similarly, there is a minimum to the product of the uncertainty of momentum Δp times the uncertainty in the position Δx of a particle that possesses that momentum; it is given by:

$$\Delta p \, \Delta x \geq \frac{1}{2} \hbar,$$

One can easily oversimplify problems involving the wave/particle duality of photons using naïve concepts. However naively, it is still safe to assume that a photon of ultraviolet light such as the Lyman-α photon emission from a hydrogen atom, involves a wave packet of about $n = 10^7$ wavelengths of the radiation. This gives a 'coherency length' δ of the photon of approximately,

$$\delta \quad = n \, \lambda \cong 10^7 \times 1.2 \times 10^{-5} = 120 \text{ centimeters.}$$

What this means is that a photon of visible light that travels at the universal speed of light in a vacuum will require a time interval Δt to pass any point as, for example, entering/leaving an atom of:

$$\Delta t \quad = \delta \, / \, c = n \, \lambda / c \quad \cong 4 \times 10^{-9} \text{ seconds}$$

However naïve, these numbers are quite descriptive of the situation.

Fourier analysis that addresses the mathematics of the duality of electromagnetic radiation demonstrates that only an infinite wave train could possibly be monochromatic (i. e., possess but a single frequency). There is accordingly a continuous broadened distribution of frequencies and wavelengths involved even in a single photon of radiation. So the photon energy given by expressions $E = h\,\nu$ or $E = h\,c\,/\,\lambda$ must be only approximate, i.e., exhibiting a corresponding variation in wavelength λ that is tied directly to the length of the wave train of the photon.

In terms of the wavelength of the radiation, we have for this uncertainty in energy ΔE, the following:

$$\Delta E = h\,\Delta\nu = c\,h\,\Delta\lambda\,/\,\lambda^2$$

Therefore, in accordance with Heisenberg's uncertainty principle, the variation in observed wavelength is:

$$(c\,h\,\Delta\lambda\,/\,\lambda^2\,) \times (n\,\lambda/c) \geq h\,/\,4\,\pi = \tfrac{1}{2}\,\hbar$$

So that,

$$\Delta\lambda \;\geq\; \lambda\,/\,4\,\pi\,n\,c^2 \;\cong\; 1.011 \times n \times 10^{-22}\ \text{cm}$$

This is indeed a very narrow width of an atomic line, but it is not infinitesimal. Refer to figure 14.2 for an illustration of the relationship between line width and number of wavelengths per wave packet.

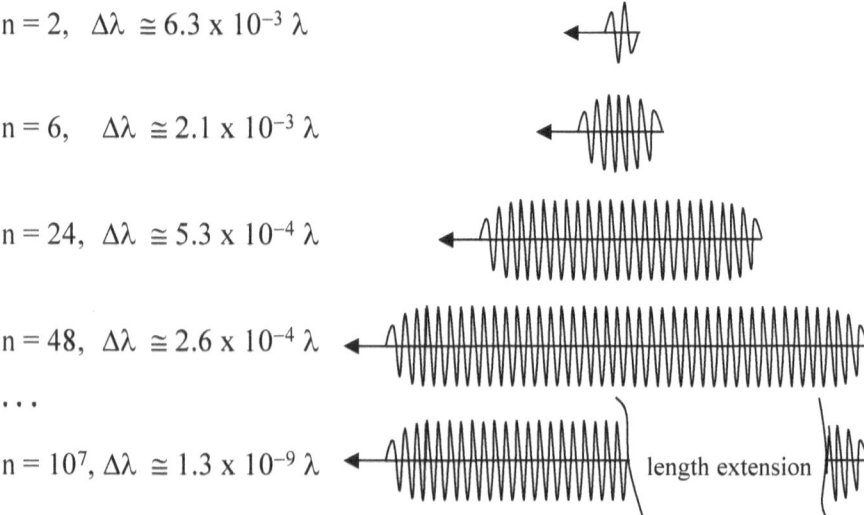

$n = 2,\quad \Delta\lambda \cong 6.3 \times 10^{-3}\,\lambda$

$n = 6,\quad \Delta\lambda \cong 2.1 \times 10^{-3}\,\lambda$

$n = 24,\quad \Delta\lambda \cong 5.3 \times 10^{-4}\,\lambda$

$n = 48,\quad \Delta\lambda \cong 2.6 \times 10^{-4}\,\lambda$

$\ldots$

$n = 10^7,\quad \Delta\lambda \cong 1.3 \times 10^{-9}\,\lambda$

length extension

Figure 14.2: Wavelength uncertainty as a function of photon length

When the effects of Doppler shifts are addressed for the myriad atoms involved in any thermal medium, there will be a continuous range of wavelengths/frequencies of atomic line radiation over a much more appreciable interval in the associated thermal radiation.

Of course the wave particle duality applies to matter as well, but because of the typically much greater mass/enery, it applies to a lesser extent for which the *complimentarity principle* applies which states that in the limit of larger entities, the more traditional classical laws pertain.

How do basic particles and photons relate to each other?

Energy and momentum of a particle are *not* directly proportional to each other. Radiational energy on the other hand *is* directly proportional to an analogous expression for momentum associated with that same radiation. There are dramatic consequences that derive from this very essential difference between the rather flexible energy and momentum relationships of particulate components of a gas and the similar (but more constrained) relationships of the radiation associated with that very same gas. This disparity is a well-known fact of which everyone, for whom it should matter, is aware. But those who have tackled this problem have typically continued to be unaccountably oblivious to the role of this essential difference in the context of analyses employed by both Boltzmann and Einstein in the development of their respective models of particles and radiation. Incompatibility of energy and momentum relationships introduces a difficult additional constraint on conservation principles employed in determining the effects of interactions that occur within thermal substances.

In analyzing the interactions using classical Newtonian mechanics the applicable functional relationship between the conserved kinetic energy $E(p)$ and the momentum p of a particle exhibits the following nonlinear squared relationship,

$$E_k (p) = p^2 / 2m_o$$

This is shown by the dotted line in figure 14.3. Here m_o is the constant (rest) mass of the submicroscopic particle independent of the particle's velocity. This results from the classical formulas for kinetic energy,

$$E_k = \frac{1}{2} m_o v^2$$

and the associated formula for momentum is

$$p = m_o v,$$

with v the relative velocity of interacting particles.

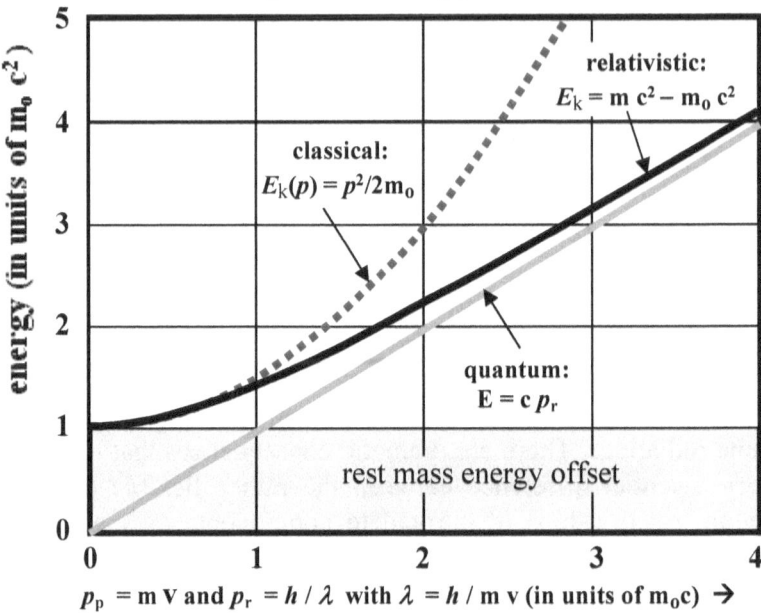

Figure 14.3: Energy and momentum relationships for particles and photons

However, Einstein's relativity theory demands a revision of these classical formulas for particulate matter with the mass of a particle now dependent on its velocity. The relevant relativistic equations become respectively:

$$E_T(v) = m(v) c^2$$

and

$$p(v) = m(v) v$$

for the energy and momentum of such a particle. And of course,

$$m(v) = m_o / (1 - v^2 / c^2)^{1/2}.$$

Here the dynamic mass m(v) is no longer a constant independent of its relative motion with respect to the observer for whom the equations apply. Now m_o represents the 'rest mass' of the particle and c is the universal constant speed of light in a vacuum. So that

$$E_k(v) = E_T(v) - E_{rm} = m(v) c^2 - m_o c^2$$

158

Here E_k is now the kinetic portion of the total energy of the particle, resulting from subtracting the 'rest mass energy' E_{rm} from the total.

Taking the binomial expansion of the relativistic expression for E_T and using only the first term in the expansion which is valid for all but quite extreme velocities so that v << c applies, we obtain approximate agreement with the classical expression:

$$E_T(v) \approx m_o (c^2 + \tfrac{1}{2} v^2) = m_o c^2 + \tfrac{1}{2} m_o v^2$$

But other than for small velocities, we no longer have the direct square relationship between momentum and kinetic energy. There is still a nonlinear relationship between energy and momentum, but now it's a little more complicated. And in fact, it is not a relationship that allows a direct analytic expression although it is a smooth function for which every value of p has an easily calculated value for the associated energy. As can be easily seen, it becomes closer and closer to the linear relationship exhibited by radiation as the relative velocity of a particle becomes closer and closer to that of the speed of light. This is shown as the dark solid line in figure 14.3.

The quantum behavior of photons of electromagnetic radiation on the other hand is constrained by an energy-to-momentum relationship that is strictly linear. This is in part because the rest mass of a photon is zero and also because of the wavelength dependence exhibited in the quantum theory of radiation where we have for the commensurate energy and momentum parameters that:

$$E_T = h c / \lambda$$

and

$$p = h / \lambda$$

Here λ is the wavelength of the radiation, c is the universal constant speed of light, and h is Plank's constant defined previously. Notice that the frequency v of such radiation is given by:

$$v = c / \lambda$$

Thus, photons (unlike their particulate component counterparts) are constrained by a precisely articulated proportionality such that:

$E(p) = c\,p$

This is shown in the light solid line superimposed on the plots of figure 14.3. Even for the proper relativistic equations, this is not the case for particulate components although, as illustrated in that figure, the differences do become much less at extremely high energies.

To understand the scope of what is meant here by "high energies", notice that the abscissa on the graph in figure 14.3 is in units of $m_0\,c$ for particle momentum so that it is nonlinear with respect to particle velocity. The tabulation of particle velocities with respect to that of light that are associated with this scale are provided in the table below.

momentum in units of $m_0\,c$	velocity relative to that of light
0.1	0.095
0.2	0.186
0.3	0.261
0.4	0.331
0.5	0.408
0.6	0.474
0.7	0.537
0.8	0.596
0.9	0.653
1.0	0.707
2.0	0.894
3.0	0.949
4.0	0.970
5.0	0.981

Compton scattering formulas

Scattering of X-radiation by electrons was shown by Compton to produce a lengthening of the wavelength of the scattered radiation in accordance with the following formula as illustrated in figure 14.4 taken from Bonn (2009):

$$\lambda_{final} - \lambda_{initial} = (h / m_e\,c)\,(1 - \cos \psi),$$

Here ψ is the angle of deflection of the photon and ϕ is dependent on the value of ψ. m_e is the rest mass of the electron; and h is Planck's constant. The basis of the preceding equation is the set of conservation constraints on energy and momentum laws as illustrated in the diagram.

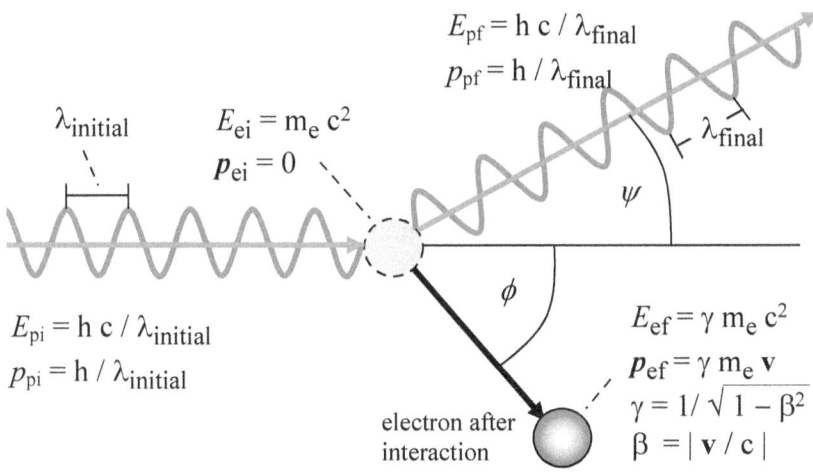

Figure 14.4: Illustration of Compton scattering conservation relations

This phenomenon does not depend on the electron being bound, of course. The resulting *change* in wavelength is *not* dependent upon the initial wavelength and has an absolute maximum value called the *Compton wavelength* defined as:

$$\lambda_C \equiv h / m_e c$$

where λ_C is approximately 0.02425 Angstroms. Expectation values of the wavelength change are very considerably less than this amount, of course; only for X-radiation and gamma rays would such quantities become significant after a single scattering. Although 'Compton scattering' typically pertains to X-ray deflection from stationary electrons, the effect applies throughout the electromagnetic spectrum as we will discuss in Part IV with regard to scattering of electromagnetic radiation.

Clearly, the lengthening of the wavelength of incident radiation associated with this (or any similar) effect must be obtained as a direct result of energy and momentum transfer from electromagnetic photons to affected material electrons as Compton demonstrated. In addition, such interactions necessarily alter the subsequent direction of travel of the photon in addition to that of affected electrons. It is inevitably associated with a 'bending' of the light path. Scattering electrons will pick up the associated loss of energy of the photon.

Is relativistic treatment required?

The obvious question is whether we need to address relativistic effects in the realm of models of submicroscopic behavior. This distinction between relativistic and classical effects might seem hardly worth mentioning with regard to the traditional applications of thermo-thermodynamics that we have discussed previously. In physical analyses a tiny second-, or higher-, order effect is frequently ignored as cancelling out when there are large numbers of entities for which the effect can be either positive or negative. Boltzmann modeled molecular behavior using classical mechanics; Einstein's relativity had not yet been invented. Later Einstein chose to use the classical Doppler formula when considering the transfer of momentum from radiation to particles. There are several reasons that suggest that may not have been a wise choice. First of all the black body radiation theory that Einstein was addressing is of considerable interest at extremely high energies for which classical formulas become inappropriate. At such high energies that are associated with high temperatures, ionization occurs which introduces free electrons whose lighter mass accommodates higher velocities and differences do indeed become appreciable as shown in figure 14.5 below.

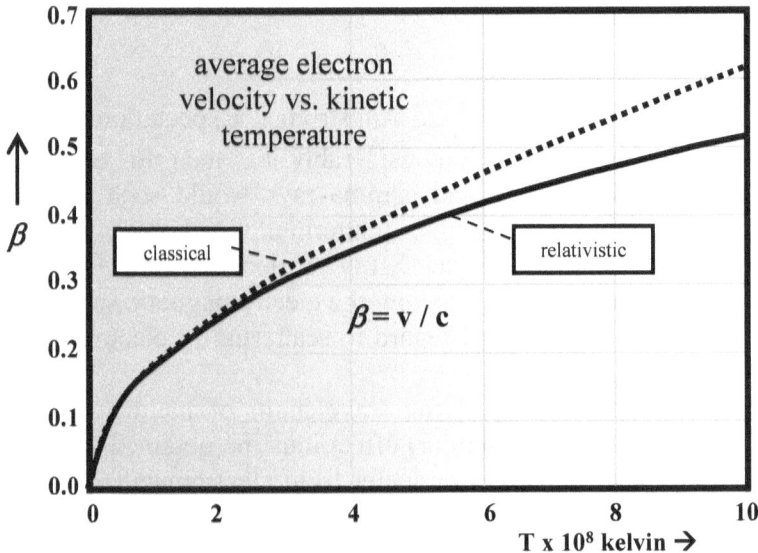

Figure 14.5: Implications of the temperature of plasma electrons on the applicability of relativistic treatment

Secondly, and significantly with regard to forward scattering, which Einstein did not address, there is a transverse Doppler side effect in relativity that is not present in the classical formula. This effect is always positive and therefore does not cancel after many interactions but grows linearly with each occurrence. Finally, even for quite modest temperatures there is a tail to the distribution of particle energies that is quite appreciable with respect to the speed of light so that the averages shown in figure 14.5 do not adequately address the range of possible velocity values that will be encountered as shown in figure 14.6.

Later we will see that the difference in the energy and momentum relationships between particle kinematics and electrodynamics produces irreversibility of interaction processes. It is therefore also associated with what we can denominate a 'cause' of irreversibility. This includes equilibrium conditions of modal number densities in the two partitions of the Maxwell-Boltzmann distribution.

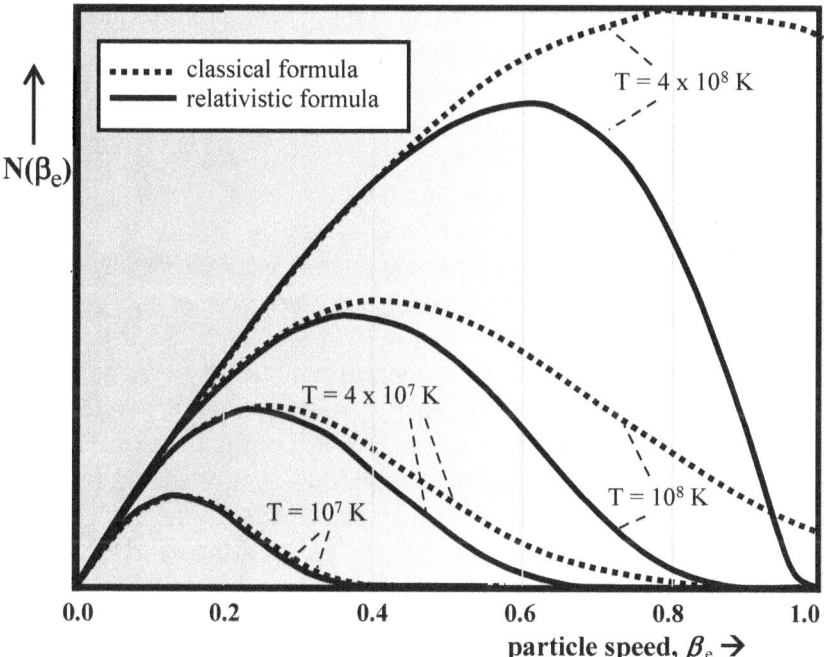

Figure 14.6: Non-normalized Maxwell-Boltzmann distribution of the electron velocities for various temperatures at equilibrium conditions

What is 'heat'?

So, what is it at the submicroscopic level that corresponds to what we call 'heat'?

We are intimately familiar with formulas for which temperature (T) is the metric for what we denominate 'heat'. We know that the value of temperature is typically assessed in terms of the macroscopically mensurable pressure (P) and volume (V) of associated encapsulated substances. For a gas in equilibrium these parameters are related by the well-established ideal gas law. However, the heralded discoveries of these relationships discussed in Part I above merely revealed empirical data readily available to anyone willing to set up experiments to determine what these relationships might happen to entail.

Pressure and volume are not the *causes* of 'heat'; they are merely correlatives of it. The same goes for their submicroscopic counterparts. Because we want a deeper understanding of nature, we seek explanations that address what it is about changes in pressure and volume that *generate or absorb* heat. What is happening to the constituents of the gas that makes it hotter or cooler? To answer that and similar questions we need to understand behavior of the lower level submicroscopic constituents of matter and the radiation that it emits and absorbs. Everyone seems to agree that there is something about the behavior of submicroscopic particles of a substance that must be the proximate cause. The following quote seems to express a consensus of that opinion:

> *"The idea underlying this concept is that in a gas, heat is nothing else than the kinetic or mechanical energy of motion of the gas molecules, so that, in expanding, a gas does work at the expense of kinetic energy of its molecules, which represent its heat energy."* [1]

The mean of the distribution of energy embodied in velocities of these submicroscopic particles is easily associated with macroscopic properties of the pressure and volume of a confined gas. But the distribution (and not just the mean) of the total energy shared by these more elementary components of a substance is of extreme significance to the stability of the macroscopic properties. A Maxwell/Boltzmann distribution of various energetic modes (and therefore also velocities) of the submicroscopic constituent molecules as a function of the macroscopic temperature provides a generally accepted working base for comprehending what is physically involved in heating or cooling, and expanding or contracting a substance.

In this way heat whose measure is temperature that is accessible by direct measurement of pressure and volume parameters using the ideal gas law is envisioned to reflect directly upon the inaccessible individual vibrational, rotational and translational velocities of constituents of the substance. That is the view of 'heat' held by most knowledgeable scientists. It is a particulate view of what constitutes heat. Erwin

---

[1] Leonard Loeb, *Kinetic Theory of Gases*, McGraw-Hill, New York, 1927 (p. 13)

Schrödinger, who master-minded breakthroughs in quantum mechanics, stated a somewhat typical position when he said:

"We know all atoms to perform all the time a completely disordered heat motion, which, so to speak, opposes itself to their orderly behavior and does not allow the events that happen between a small number of atoms to enroll themselves according to any recognizable laws. Only in the cooperation of an enormously large number of atoms do statistical laws begin to operate and control the behavior of these assemblies with an accuracy increasing as the number of atoms involved increases. It is in that way that the events acquire truly orderly features."[2]

This fallacious argument that is shared by many of the greatest minds of science, maintains in essence that the scientific reductionist agenda is an illusion – that there are indeed 'emergent' phenomena that pertain only at a particular level of complexity and (importantly) have no counterpart at lower levels of existence. A significant flaw in his argument involves the presumption without supporting evidence or explanation that "completely disordered heat motion" is comprised of something *other than*, and presumably *opposed* and *superior* to, the typical motions associated with the "events that happen between a small number of atoms." This has certainly never been established, nor could it be. And furthermore, it totally opposes assumptions of the kinetic theory of gases that most nearly account for a wide range of related phenomena.

Certainly the equipartitioning of energy among various constituents and modes of behavior would be completed to a very high degree with a relatively small number of constituents. If, in fact, "heat motions" were to constitute a uniquely irreversible behavior pattern counter to the otherwise understood reversible processes in which smaller numbers of atoms "enroll themselves", the statistical treatment of the latter reversible processes could *not* suffice as the explanation of the former. Significantly, statistics most certainly do *not* "operate and control the behavior" of an associated assembly; statistics merely describe it with more or less distortion associated with whatever data compression characterizes the statistical approach. This was the essence of a criticism by Lochschmidt that Boltzmann had taken to heart eighty years earlier.

---

2 Erwin Schrödinger, *What is Life?* Doubleday, New York, 1956.

Furthermore, Boltzmann's collision analyses do not demonstrate a method by which one distribution of velocities evolves into another. They demonstrate that a Maxwell/Boltzmann distribution of velocities is stable. That is also what statistical mechanics demonstrates.

A more complete answer

But there is another view that we will propound further on, for which Planck's black body distribution of radiation becomes paramount. Of course both the particulate and radiational distributions within a single substance that is in equilibrium must be compatible as Einstein so ably demonstrated. However, despite the equipartitioning of total energy, these two energy distributions are certainly not the *same* distribution. Therefore, to identify heat as uniquely tied to an archaic classical physics view of one rather than the other, to ignore a combination of both, as the much more essential aspect would be, and has been, a serious mistake. Also, to assume that relativistic and photoelectric effects would have little bearing on the interactions of photons and high speed particles has been a serious mistake. Even Einstein neglected relativistic effects in his definitive treatment of the quantum theory of electromagnetic radiation.

Understanding the role and subtleties of the operation of radiant energy in processes resulting in the equal partitioning of the various energetic modes within the total energy of an ensemble of particulate matter is of paramount significance to any proper understanding of heat and/or entropy. The authors maintain, however, that any attempt to explain or try to understand the nature of heat based strictly on a calculation presuming mechanical equivalence, whether incorporating statistical treatment or not, is incorrect and misleading from an epistemological perspective. It has certainly befuddled the current understanding of cosmology

Unlike particle kinetic energy, radiant energy is such that a contained volume of a heated substance immersed in a total vacuum will transmit heat into and across material boundaries and across the vacuous expanse surrounding these boundaries to remote regions of the environment. In this process the substance is cooled unless surrounded by remote regions containing substances at the same or a higher temperature. Since this would be accomplished with no intermediate material particle interactions when a container is placed in a vacuum, heat must necessarily be different from, or at least more than, *mere* random motions of constituent material particles. The relevance of radiant aspects of heat also becomes apparent by noting the unique relationship of the Planck distribution of energies associated with the internal radiations of a heated substance.

This is in addition to, and different than, the relationship characteristic of the Maxwell-Boltzmann distribution of mechanical particle energies applying between constituent particle motions and temperature. Einstein found the relationship between these two unique equilibrium distributions by noting that Planck's distribution is the *only* distribution of radiant energy frequencies compatible with maintaining a Maxwell-Boltzmann distribution of mechanical energies of the constituent particle at the same temperature.

However, even this does not provide a model of heat, nor yet does it address the more essential issue of the interactive mechanism involved in irreversibly driving closed systems to equilibrium, nor yet the absorption and dissipation of heat by open systems. We have been left without answers concerning the specific nature of the association of heat with the mechanical motions of constituents of systems of material particles and this lacuna is essential to the even more significant gaps in our current understanding of the profound issues surrounding entropy.

Equipartitioning of total internal energy of mechanical systems at equilibrium seems to be an inevitable consequence of the physical processes affecting the interactions of constituents. These interactions have been known for over a century to be mediated by electromagnetic photons. But the processes by which they occur and an associated 'thermalization' process have never been adequately explained. We will show that this mediation process is *not* reversible as has been commonly, but erroneously, assumed without proof. Nor is the insufficiently-understood thermalization of radiation a reversible processes for similar reasons. These processes ultimately bring both the Maxwell-Boltzmann distribution of particulate energies and the Planck distribution of radiational energies into a stable relationship much as Einstein showed a century ago. But significantly in addition, they introduce irreversible consequences into the system.

What the authors will demonstrate is that the introduction of photon mediation with its associated transverse Doppler effect, an extremely tiny, but unilaterally all of the same sign, side effect occurs on every interaction. The total effect is *not* time (or velocity) reversible. Thus, because the effect on an individual mediated interaction or any countable number of them would be immeasurable for systems at temperatures for which velocities are much less than the speed of light, it is inevitable that any measurable loss of energy or irreversibility effect would only be appreciable at the macroscopic level where astronomical numbers of interactions are involved and so they must be treated by statistics.

\

15: Einstein's Quantum Theory of Radiation

"But if these hypotheses on the interaction between radiation and matter turn out to be justified, they must produce rather more than just the correct statistical distribution of the internal energy of the molecules: for there is also a momentum transfer associated with the emission and absorption of radiation; this produces, purely through the interaction between radiation and the molecules, a certain velocity distribution for the latter. This must evidently be identical with the velocity distribution of the molecules which is entirely due to their collisions among themselves, i.e. it must agree with the Maxwell distribution."[4]

In Einstein's analysis which gave rise to the quantum theory of radiation, two atoms that exchange a photon do so in a very constrained and (at first sight) reversible process not unlike a child's seesaw or teeter-totter as shown in figure 15.1.a. Quanta of the internal electronic potential energy of electrons in 'higher' energy states are transmitted to other atoms in lower states via electromagnetic radiation whose frequencies embody the discrete amounts of energy transmitted between such atoms as shown in figure 15.1.b. Interestingly Einstein was able to derive the Planck distribution of radiation from just this level of analysis, although he was motivated to wonder about the impact of momentum exchanges that must accompany any such energy exchange. By exploring that aspect he was able to also determine the

---

[4] Albert Einstein, "On the Quantum Theory of Radiation," *Sources of Quantum Mechanics*, Dover, New York, 1967. (p. 64)

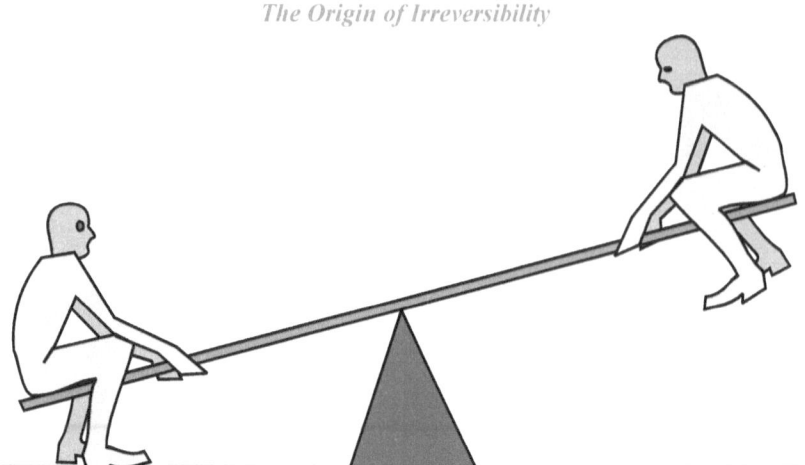

(a) principle of precise dynamic balance

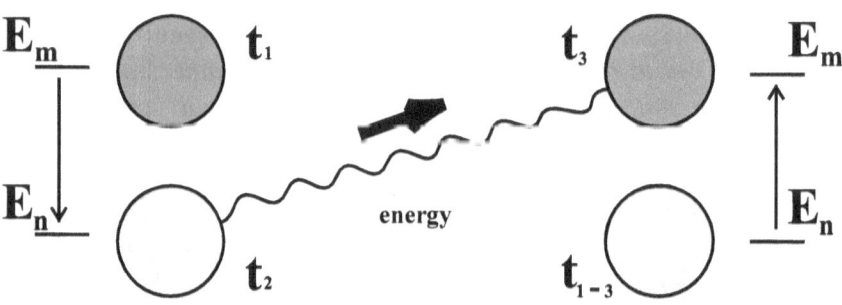

(b) energy considerations

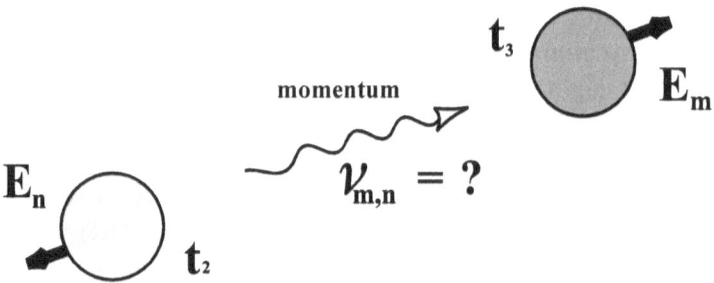

(c) momentum considerations

Figure 15.1: Precise dynamic balance of conservation laws

additional necessary compatibility requirements that must apply between the work of Maxwell and Boltzmann, and that of Planck and his own.

There is more to be explored here than what was pursued by Einstein in this regard however. It turns out to be typical of the exchanges of energy between atomic particles that they are mediated by photons. This certainly adds considerably to the interactions that were formerly treated as exclusively the domain of 'elastic' collisions between molecules that we explored in previous chapters. Interestingly the momentum and energy of a photon are both linearly related via the frequency of the radiation as we have discussed, whereas mechanical kinetic energy and momentum are *not* linearly related in classical physics. Nor for that matter is there any more compatibility in the corresponding relativistic treatment:

$$E_R \sim p = (h \nu / c)$$ *radiational relationship*

$$E_M \sim p^2 = (m v)^2$$ *mechanical relationship*

In the proportionality statements above, symbols E and p refer to energy and momentum, respectively, while subscripts R and M refer respectively to radiant and mechanical entities. This simple difference makes it necessary that the photons of electromagnetic radiation, when treated as particles, possess only the single possible speed, c, *the speed of light* in vacuum, and zero *rest mass* energy. It also complicates the precise dynamic balance to be maintained when the radiation interacts with mechanical aspects of a system that is mutually constrained by conservation laws with regard to the total energy and momentum.

If one limits consideration to the conservation of energy, any exchange of radiation would appear trivially reversible just as in the teeter-totter shown in figure 15.1.a. Identical photons could be exchanged back and forth between atoms *ad infinitum* with nothing significant changing in the least. However, in consideration of the dynamics of those same two interacting atoms or molecules, the picture changes dramatically if the effects of the exchanged momentum are to be taken into account as well. See as an analogous illustration the situation of two ball players on skate boards on a frictionless playing field that is shown in figure 16.1 in the next chapter. Clearly the initial interaction situation cannot be perpetuated by velocity or time reversal.

Einstein considered the case where two involved atoms or molecules are relatively at rest similar to the ball players at the top of figure 16.1. However even in this case momentum is exchanged as will be shown, and more significantly, Doppler effects alter effectiveness of the *precise dynamic balance* on which his analysis depended. Therefore, in general the amount of energy emitted by the one atom or molecule is no longer unconstrained in its delivery to the other when relative velocity that inevitably alters the frequency of the radiation is taken into account. Furthermore, to assess the impact of this using conservation laws, one must take into account associated relativistic effects, which uncharacteristically Einstein did not do in his analyses.

Einstein's insight

Einstein's extreme success in discovering the long sought rational explanation for Planck's empirical radiation density formula resulted in the quantum theory of radiation that has retained essentially the same form ever since. There are many aspects of note in Einstein's seminal paper, but certainly classification of two types of emission, i.e., *stimulated* and *spontaneous*, had profound significance on later technological developments in the twentieth century.

Stimulated emission involves an implicitly *closed* system involving 'induced' radiation that seems to be *caused by* the ambient internal radiation density. *Spontaneous* emission on the other hand is treated as having *no* identifiable physical cause at all and is essentially free to exit an *open* system. The treatment of spontaneously emitted radiation is very similar to that of random decay phenomena including the decay of submicroscopic particles.

It must be noted that Einstein's treatment addresses emission and absorption of radiation exclusively. Most notable is the recognition of the fact that these phenomena involve discrete quanta of energy exchanges between particulate and radiational energy as we showed above. He accomplished this by applying the emerging understanding of there being discrete quantum energy states $Z_1, Z_2, Z_3,... Z_n,...$ associated with internal energies $E_1, E_2, E_3, ...E_n,...$ in the constituent atoms/molecules of any system. Maxwell's and Boltzmann's work had made it clear that the probability of a constituent element in an equipartitioned system in a state Z_n is:

$$F(E_n) = p_n\, e^{-E_n/kT}$$

Here the parameter p_n is independent of the temperature T. It is characteristic of the particular type of atom/molecule and the particular allowable quantum state Z_n of that atom/molecule.

In a manner reminiscent of Boltzmann's treatment of molecular energy exchanges, Einstein considered individual transitions between states of the atoms/molecules as the mechanism of distributing energy among constituents of the system. In this analysis he considered associated photons of radiation as the bearers of the transported energy as shown in frame 2 of figure 15.1 above.

quantum phenomena involved in energy transitions

In hydrogen gas, photons with frequencies in the vicinity of atomic spectral absorption lines of the element are absorbed in transitioning atoms to their higher energy levels. This phenomenon, perhaps best described by 'the Bohr atom', involves quantum energy transitions of electrons captured in the coulomb potential energy wells of the proton nucleus of the hydrogen atom. In this conceptualization shown in figure 15.2, bound electrons are restricted to allowed orbital situations for which their energies are limited to specific quantum energy levels:

$$E_n = - m_h e^4 / 4\pi \hbar^2 n^2 = - 2.18 \ (1/n^2) \ x \ 10^{-11} \ ergs = - 13.6 / n^2 \ eV$$

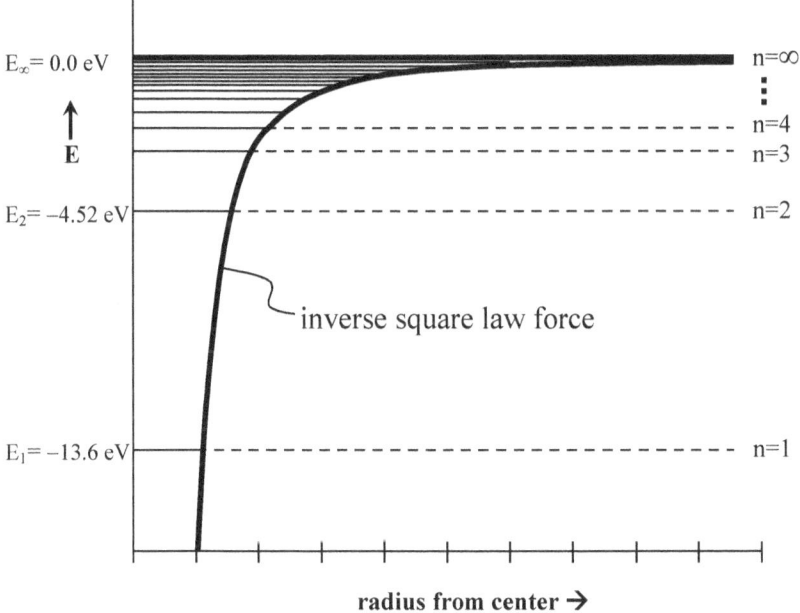

Figure 15.2: Potential energy well and quantum states of the hydrogen atom

where n is an integer n = 1, 2, 3, ...,∞, m_h = 1.6726 x 10^{-24} gm is the mass of the hydrogen atom, and e is the electronic charge. The minus sign has to do with these energies being associated with bound states in a potential energy 'well'. We also have Planck's constant, which as we noted before is: $h = 6.626 \times 10^{-27}$ erg seconds. And in addition we have the conversion constant for converting energy units from ergs to electron volts, where,

$$1 \text{ eV} = 1.60 \times 10^{-12} \text{ ergs}$$

The important Lyman energy transition series is labeled in figure 15.3 along with the Balmer and Paschen series that play somewhat lesser roles in usual discussions.

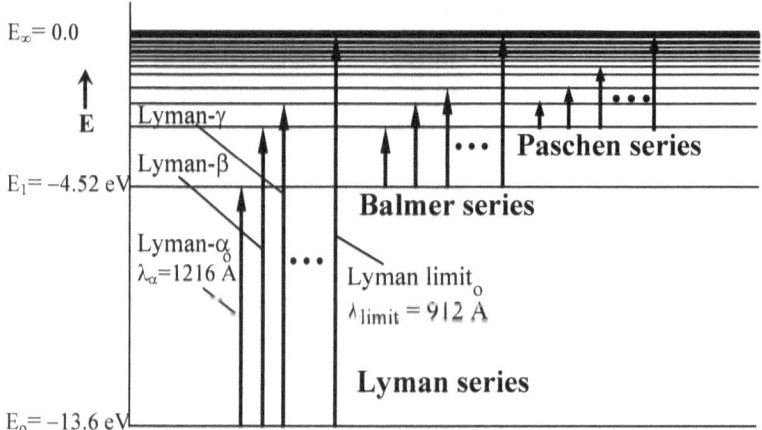

Figure 15.3: Hydrogen emission/absorption line series

If the hydrogen atom is in its lowest energy state E_o with a single electron in its lowest 'orbit', the atom can transition to the next higher energy state E_1 by absorbing a photon whose energy is given by,

$$\Delta E_\alpha \equiv E_1 - E_0 = h\,c\,/\,\lambda_\alpha \cong 1.65 \times 10^{-11} \text{ ergs}$$

This is the energy exchange in producing the 'Lyman-alpha' transition.

A *Principle of Detailed Dynamic Balance* resulting from Einstein's work involves the discrete energy transitions between and among the various energy levels of the constituent atoms being precisely constrained to continuously maintain conservation laws as was shown in figure 15.1. Thus, even though one atom or molecule can emit radiation to transition a state from say Z_{10} down to Z_7, for example, this energy transition cannot be *precisely* balanced by three atoms or molecules, each absorbing portions of the photon to effect the following transitions: Z_7 to Z_8, Z_8 to Z_9, and Z_9 to Z_{10}, in three such molecules respectively or in three separate stages of a single molecule. This restriction applies in both directions. *Photons are indivisible.* The conservation of energy is maintained on *each and every* transition indivisibly as shown in figure 15.4. In the figure the potential energy well of a hydrogen atom and its quantized energy levels for its bound electron are shown. In this process there is the closed system syndrome that accommodates reversibility.

potential energy wells for

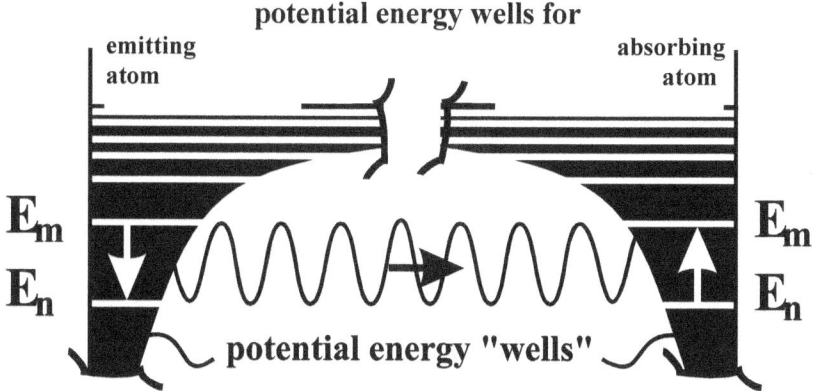

Figure 15.4: The precise potential energy balance of exchanges of quanta of energy between relatively stationary atoms

Einstein modeled absorption as a resonance phenomenon *caused* by incident radiation. Accordingly he hypothesized that for an atom in a lower state Z_n to transition to the higher energy state Z_m, it must absorb a photon of frequency $v_{m,n}$, where:

$$(E_m - E_n) = h \, v_{m,n}$$

The probability of the particular frequency $v_{m,n}$ being available for absorption in the current ambient radiation distribution depends upon the radiation density $\rho(v_{m,n})$. Thus we arrive at the differential probability of absorption, dP_{BA} during the time interval dt as follows:

$$dP_{BA} = B^m_n \, \rho(v_{m,n}) \, dt \qquad \textit{stimulated absorption probability}$$

Here B^m_n is a constant that is independent of time. This equation was rather an expected extension of earlier understandings of what was involved in these electromagnetic processes. But Einstein incorporated an additional hypothesis that has had many far-reaching consequences including laser technology. The hypothesis is that a radiation field $\rho(v_{m,n})$ also stimulates atoms in the Z_m state to *emit* radiation and transition those atoms to the lower state Z_n in so doing:

$$dP_{BE} = B^n_m \, \rho(v_{m,n}) \, dt \qquad\qquad \textit{stimulated emission probability}$$

Here B^n_m differs from B^m_n but it also is a constant independent of time.

In addition to this bold conjecture, he additionally hypothesized the existence of two unique types of emission processes. With regard to what he defined as *spontaneous* emissions of radiation, he modeled the probability of such emissions that transition an atom from state Z_m to state Z_n with $E_m > E_n$ during an interval of time dt very simply as follows:

$$dP_{AE} = A^n_m \, dt \qquad\qquad \textit{spontaneous emission probability}$$

In these equations, A^n_m, B^m_n and B^n_m are all constants independent of time but with unique values for each unique combination of the pair of indices, m and n. Collectively these three constants are denominated 'the Einstein coefficients'. A^n_m is sometimes separately referred to as either a 'transition probability' or the 'coefficient of spontaneous emission'.[5]

Having thus defined all of the allowed transitions of energy in his model, he proceeded to define a stable distribution of radiation $\rho(v)$ as that which maintains the Maxwell-Boltzmann distribution of molecular energy states $f(E_n) \sim e^{-E_n/kT}$. To this end he stipulated that:

$$p_n \, e^{(-E_n/kT)} B^m_n \, \rho(v_{m,n}) = p_m \, e^{(-E_m/kT)} (B^n_m \, \rho(v_{m,n}) + A^n_m)$$

to insure equal probability of emission and absorption between any two energy levels to conserve energy. He added the constraint that as the radiation density tends toward infinity the temperature must tend toward infinity as well. This is a straight-forward constraint that was to be expected. From it we must have that:

$$p_n \, B^m_n = p_m \, B^n_m$$

Plugging this equality into the previous equation, we obtain:

---

[5] R. W. Ditchburn, *Light*, 2nd Ed., Interscience, New York, 1963 (pp. 681-684)

$$\rho(v_{m,n}) = (A^n_m / B^n_m) /(e^{((v_{m \cdot n})/kT)} - 1)$$

Thus we obtain Planck's distribution of thermal radiation as a consequence of imposing dynamical equilibrium.

Of course, to match Planck's empirical formula and the previous work of Rayleigh and Jeans discussed in chapter 7, we must have that:

$$(A^n_m / B^n_m) = 8 \pi h (v_{m,n} / c)^3$$

Thus, we get the formula dicussed earlier, which is plotted in figure 15.5 where radiation frequency is employed as the abscissa.

Clearly, for visible light (lower than 10^{15} Hz frequencies) the ratio shown above remains extremely small so that only for very low radiation densities (involving low temperatures or very high frequencies at which the Planck factor would nullify the effect) would *spontaneous* emission predominate.

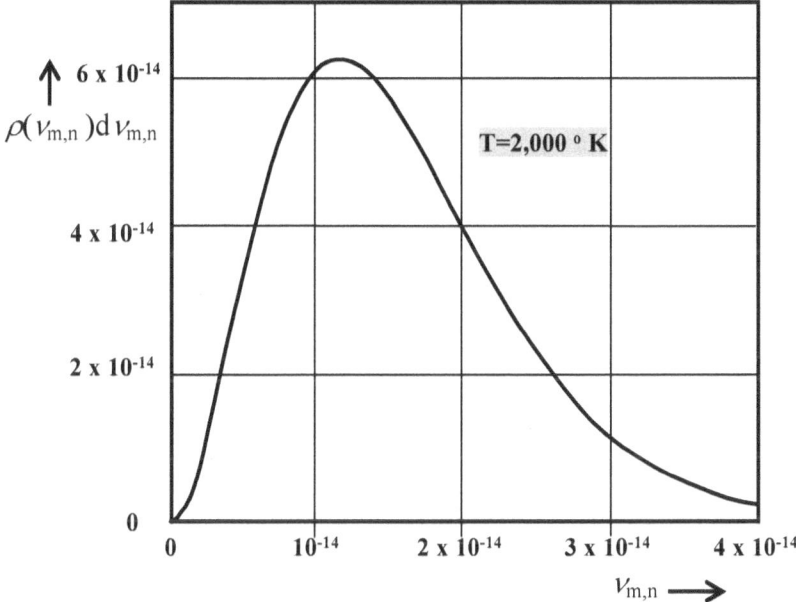

Figure 15.5: Blackbody radiation frequency distribution

extending Einstein's model

Einstein's analysis with regard to deriving the Planck distribution of blackbody radiation is ingenious of course, but there seems to be an egregious flaw in it as well. It had previously been presumed that all emission of radiation was of a 'spontaneous' *a causal* type until Einstein's definitive treatment of the *second* 'stimulated' emission type which actually predominates in thermal systems. But other than somewhat vague analogies in the biological sciences, nature seems to abhor duplicity. So why would there be *two* ways for an atom to emit radiation of *exactly the same* kind? Clearly, 'God would *not* have done it that way' to borrow phraseology from Einstein himself. So why does the presumption work so well?

Consider that Einstein's analyses apply exclusively for the idealized situation that characterizes a *blackbody* thermal source of radiation: In blackbody cavity radiation the emission and absorption coefficients are precisely equal. By assuming that all internal emissions are absorbed internally with 'spontaneous' emissions not balanced by a receptive molecule, he is in essence assuming that 'spontaneous' absorption must be included in $p_n B^m_n$ to account for absorption of the spontaneously emitted photons. Let us separate out the requisite unique type of absorption by defining a term $p'_n A^m_n$ balance emission and absorption of the 'stimulated' type. For this to work we must have that,

$$p'_n A'''_n = p'_m A''_m \qquad \text{\textit{spontaneous absorption probability}}$$

where $p'_m A^n_m$ is an emission term not included in Einstein's model.

In figure 15.5 (below) Einstein's model shown earlier as figure 10.2 has been generalized to include the environmental setting of a thermal source/sink of radiation. Only at equilibrium with its environment would the previous condition be met by a thermal system. Only for that specific condition does Einstein's analysis apply. The internal radiation density that Einstein derives is actually determined by A^m_n, *not* A^n_m. These two coefficients just *happen* to produce equalized effects when the system under study is in equilibrium with its environment. And if they are not equal, then we have that $p_n B^m_n$ is *not* equal to $p_m B^n_m$ and heating or cooling must be occurring.

We must, therefore, extend this model to incorporate the interface to the environment as shown in figure 15.6. In doing this, it will become apparent that *spontaneous* emission is not a totally unique category of radiation emission. It is emission *stimulated* by a radiation field from *outside* the otherwise modeled behavior. A^n_m and A^m_n pertain to emission and absorption of radiation by the system as a whole in an *external* radiation field.

In addition to determining the internal radiation density based on the presumed Maxwell/Boltzmann distribution of *excited* energy states of the molecules, Einstein provided analyses to demonstrate how energy exchanges between two very diverse energetic modes can be effected. As a part of his analyses he showed that, by affecting momentum associated with what had previously been considered to be purely mechanical motions of the molecules, photon impacts would disrupt their equilibrium distribution unless the impacts produced effects that were themselves so distributed.

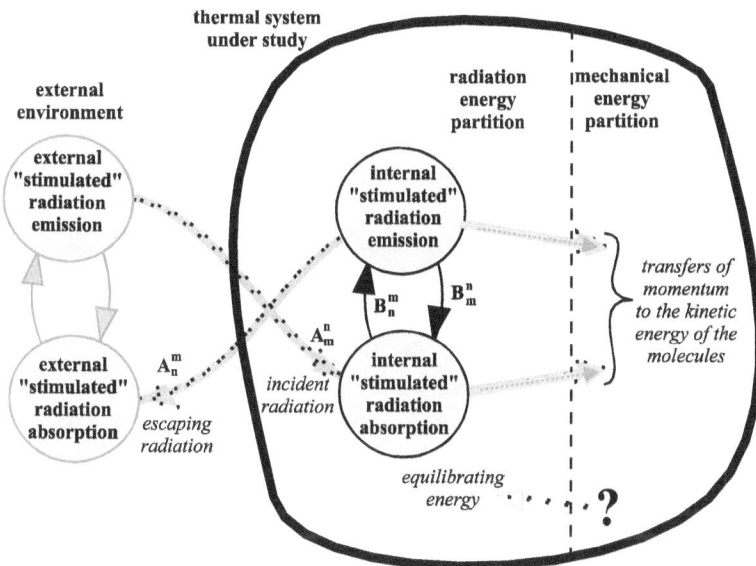

Figure 15.6: Generalized model of thermal emission/absorption applicable beyond idealization of a blackbody radiator

This must therefore constitute the means whereby, radiation energy is transferred to mechanical motions of the molecules to achieve equilibrium as required by the principle of equipartitioning among all the various (and highly varied) energetic modes of a given system. In this case, the exchanged radiation would typically increase the velocities of emitting and absorbing molecules in the distribution if the mechanical motions were associated with a lower temperature than that of the ambient radiation. Therefore also, if one were to irradiate a gas with radiation characterized by a Planck distribution of a higher temperature than that associated with molecular motions of the gas particles themselves, those motions would thereby gradually increase till they had attained a Maxwell/Boltzmann distribution compatible with temperatures of the incident radiation.

That much is evident from Einstein's original work. Obviously he recognized this compatibility as a most important condition on the distribution of radiation, but it is not clear that he realized how this fits into a more general model of which it is but one conceptual component. In particular, *how is the equipartition of energy between the two distributions of disparate energetic modes effected in toto*? In his case it is momentum of photons that are transferred to the mechanical motions of the involved molecules, increasing their energy if there happened to be less translational energy than ambient radiant energy.

But what happens when the molecules happen to have a total mechanical energy in excess of that currently characterizing ambient radiation? Again, it is straight-forward to conjecture that resulting transfers of momentum to mechanical modes of the molecules might, by falling short of expectation amounts, reduce the velocity profile in the otherwise stable distribution of velocities and perhaps thereby lower the temperature. Einstein must have been satisfied that *that* was indeed the rest of the picture. But it is *not* the rest of the picture, since it cannot explain how the radiation field associated with molecular motions comes into existence in the first place.

We will proceed to complete the model of microscopic interaction behavior and show how once again the concepts of 'modern' physics are at the heart of equilibrating energy between mechanical and other energetic modes – and is responsible for irreversibility

16: Mediated interactions

"It is clear that in an assembly of atoms and molecules, equilibrium is dynamic rather than static. All the time some atoms are absorbing radiation and some atoms are emitting radiation. These absorptions and emissions are accompanied by changes in the stationary states of the atoms, so that the conservation laws are satisfied in each individual process."[2]

Of course in all cases radiation is a byproduct of one sort of interaction or another between material substances. Whereas neutrally charged molecules can interact more or less as earlier chapters describe, photons cannot interact with each other at all. Thermal radiation does involve ensembles of particulate matter like the molecules studied by Boltzmann; however, that treatment must be augmented to include photon exchanges between molecules for which the inevitable 'middle men' in these interactions are photons. Rather than exclusively the result of elastic collisions, the effects of these mediated interactions contribute substantially to the energy distribution of thermodynamic systems. However, these interactions between molecules are not typically reversible; they are the source of irreversibility and entropy.

---

[2] R. W. Ditchburn, *Light*, 2nd Ed., Interscience, New York, 1963 (p. 681)

a mundane example of mediated interactions

What can transpire between molecules in a gas is somewhat similar to a game of pitch and catch on skateboards as illustrated in figure 16.1. In this illustration energy is exchanged between the players without direct contact. Because there is also an exchange of momentum, if the game is prolonged the players will ineluctably drift apart. But a direct reversal of velocities would restore their altered status as we will show.

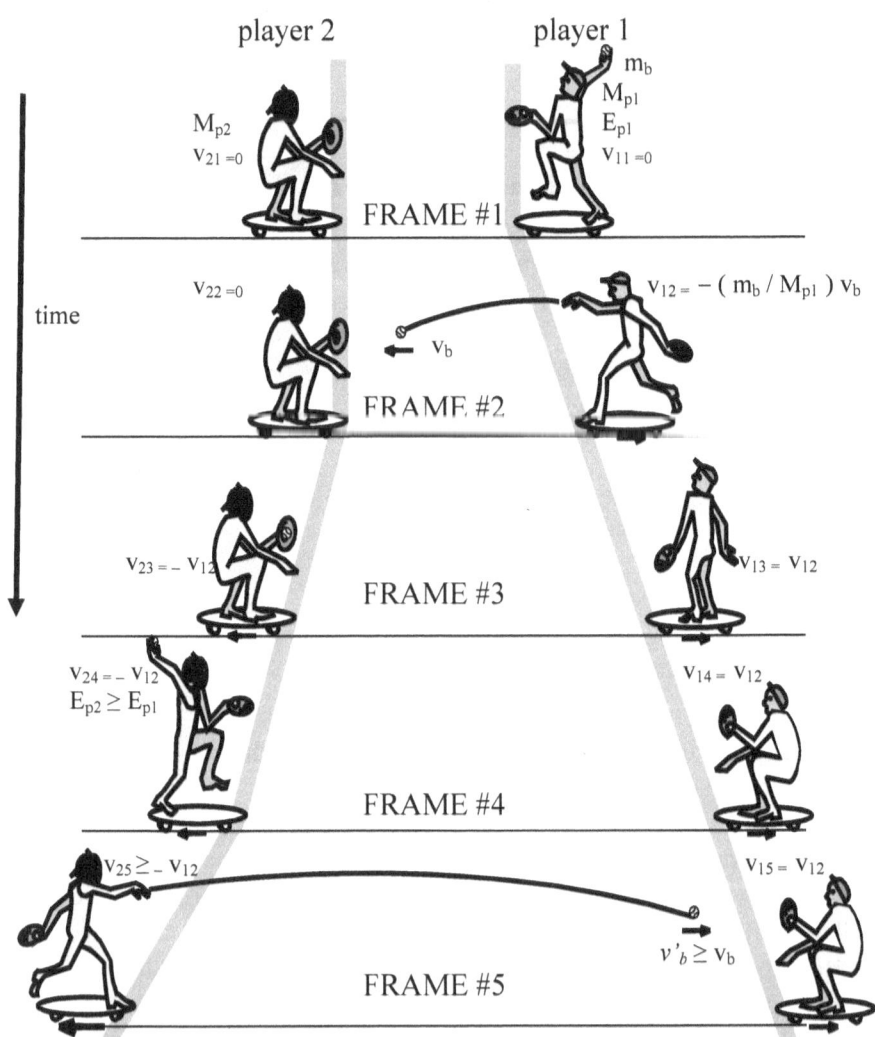

player 2   player 1
m_b
M_{p2} M_{p1}
$V_{21} = 0$ E_{p1}
                             $V_{11} = 0$

FRAME #1

$V_{22} = 0$ $V_{12} = -(m_b / M_{p1}) v_b$
time v_b

FRAME #2

$V_{23} = -V_{12}$ $V_{13} = V_{12}$

FRAME #3

$V_{24} = -V_{12}$ $V_{14} = V_{12}$
$E_{p2} \geq E_{p1}$

FRAME #4

$V_{25} \geq -V_{12}$ $V_{15} = V_{12}$
                             $v'_b \geq v_b$

FRAME #5

Figure 16.1: The dynamics of a 'catch' sequence on skateboards

In the series of exchanges that are depicted in figure 16.1 the total momentum is conserved and equal to zero in each frame. Given the amount of energy exchanged by the ball, conservation of momentum will determine the resulting motion of the two players in each case.

But what about the conservation of energy? How is that to be handled when there is no initial translational motion of the two players and therefore no initial kinetic energy, but there is thereafter? The potential energy in a cocked spring-loaded device set to release the ball initially would suffice as a source of potential energy that would be conserved thereafter. One must suppose that the energy possessed by ballplayers also provides a similar potential (latent) energy sufficient to balance the conservation of energy equation between frames.

The amount of 'potential energy' E_{p1} to be released by player 1 must be equal to the total of the kinetic energy of the ball after it is thrown plus the kinetic energy vested in the equal and opposite reaction of player 1 after tossing the ball. So to demonstrate the conservation laws between frames according to classical physics, we have:

FRAME #1
In frame #1 both players whose masses are $M_{p1} = M_{p2}$ are essentially stationary on their skate boards as player 1 winds up to throw the ball of mass m_b.

$$p_{total} = 0$$

$$E_{total} = E_{p1}$$

FRAME #2
In frame #2 player 1 throws the ball and thereby takes on the equal and opposite momentum and kinetic energy of the thrown ball, so

$$p_{total} = m_b v_b + M_{p1} v_{12}$$

$$E_{total} = \tfrac{1}{2} m_b v_b^2 + \tfrac{1}{2} M_{p1} v_{12}^2$$

From the conservation of momentum we obtain, $v_{12} = - (m_b / M_{p1}) v_b$ and therefore,

$$E_{p1} \geq m_b v_b^2 (1 + m_b / M_{p1})$$

FRAME #3
In frame #3 player 2 takes on the momentum $m_b v_b$ and kinetic energy $\tfrac{1}{2} m_b v_b^2$ of the thrown ball, so that $\tfrac{1}{2} M_{21} v_{23}^2 = \tfrac{1}{2} m_b v_b^2 = - \tfrac{1}{2} M_{p1} v_{12}^2$. Thus, we obtain: $v_{23} = - v_{12}$.

FRAME #4

In frame #4 player 2 winds up to throw the ball, demonstrating that he too has an amount of potential energy similar to that demonstrated by player 1. This is in addition to his acquired kinetic energy. However, in order for this game of catch to proceed indefinitely we must have that $E_{p2} \geq E_{p1}$ since he must throw the ball further (i.e., faster).

FRAME #5

In frame #5 player 2 throws the ball, thereby taking on the additional equal and opposite momentum and kinetic energy of the once again thrown ball. If his potential energy released in throwing the ball is the same as that of player 1, the momentum and energy of the thrown ball with regard to a stationary frame of reference (certainly with regard to player 1) is less than for player 1 in frame #2 because the velocity v'_b will now be relative to the moving frame of reference of player 2.

What must be concluded from this example sequence is that to maintain a continuing interaction of this type, more and more energy must be contributed from the potential energy vested in the ballplayers. As they get separated further and further and are receding at higher and higher velocities they must throw the ball harder and harder. That is an important issue related to what we must consider in the mediation of interactions between submicroscopic material particles as well.

It is significant that in figure 16.1 time reversibility applies. If the velocities are reversed the results will be the situation depicted in the just preceding frame. This is true all the way back to the first frame where the ball players had no kinetic energy. Furthermore, if the reversal were to continue from that point to even further back in time, the results will be the same as was depicted going forward in time without time having been reversed. This is the essence of arguments that irreversibility cannot derive from the reversible laws of physics. But that is only true of the classical Newtonian laws of physics.

binding energy of atomic structure

Importantly, photons are *not* directly analogous to baseballs thrown back and forth by ballplayers and cannot be treated in exactly the same or even a similar way. As was discussed in an earlier chapter, the relationship of the conservation laws that apply to radiation differ considerably from those that apply to any material objects such as the baseball in figure 16.1. So the *before* and *after* energy and momentum situations of emitting and absorbing molecules in a gas will differ accordingly.

As illustrated by the example of mediated interactions using material objects, there is energy other than the kinetic energy of linear motion to be taken into account. There is a latent form of energy involved whenever interactions are not strictly 'elastic', i. e., objects ricocheting off each other as exclusively addressed by Boltzmann. After each player catches the ball there is, in addition to his kinetic energy, a latent form of energy that enables him to throw the ball again.

Let us consider an inanimate example to similar (but somewhat different) effect. Suppose we have two hydrogen atoms. We will not get into much detail on the structure of the atom at this point; suffice it to say that the positively charged nucleus (a proton of the hydrogen atom in the case) possesses an inverse squared potential energy well surrounding it. A negatively charged electron is 'attracted' by an associated force and becomes bound in an orbit about the nucleus. But because of the quantization of energy the electron cannot just collapse into the nucleus giving off energy continuously; there are in fact allowed energy level restrictions that come to bear as shown in figure 16.2 below. If the electron 'falls' to a lower *allowed* energy, a photon of radiation with a discrete frequency will be emitted. The energy in this photon is matched to the energy difference between energy levels in the atom. The photon's energy will be equal to the difference in the energy of the levels, ΔE as shown in the diagrams of figure 16.2.

In this figure we illustrate the inverse square potential energy well surrounding the proton nuclei of the two hydrogen atoms with their discrete energy levels for a single bound electron. If the electron is at the bottom energy level, absorption of a photon of just the right energy will transition it to a higher level as illustrated for hydrogen atom 2.

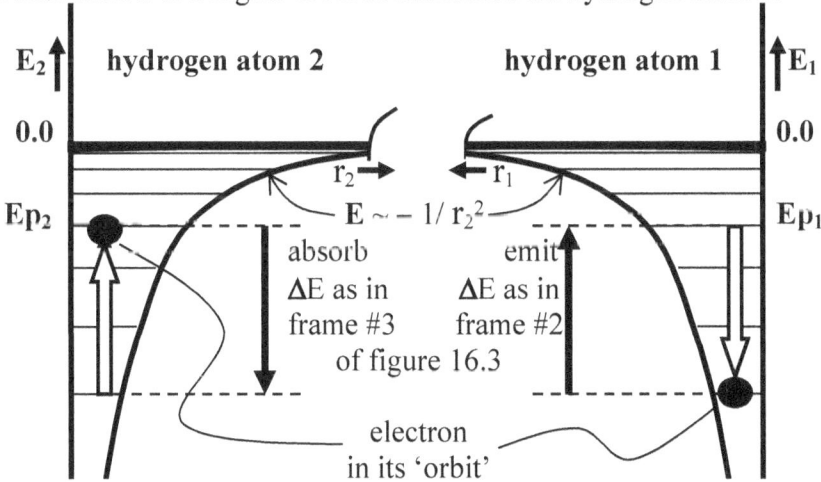

Figure 16.2: The energy exchange between two hydrogen atoms

conservation laws involving photon emission/absorption

Now suppose that in analogy to the situation depicted in figure 16.1 a photon is emitted by an atom that is ultimately to be absorbed by another. We will formulate the conservation laws so as to define all the pertinent parameters and their relationships. These relationships will prove useful even in determining whether an interaction is viable. In figure 16.3 and in diagrams on the next few pages we assume the values of the rest mass m_o (or equivalently, the rest energy $m_o c^2$) of the lowest energy state of an atom or molecule involved in the interaction is given as well as the frequency of the radiation. The amount of energy ΔE invested in the radiation that is associated with the emission or absorption of a photon is given by the expression:

$$\Delta E_{photon} = h\,\nu$$

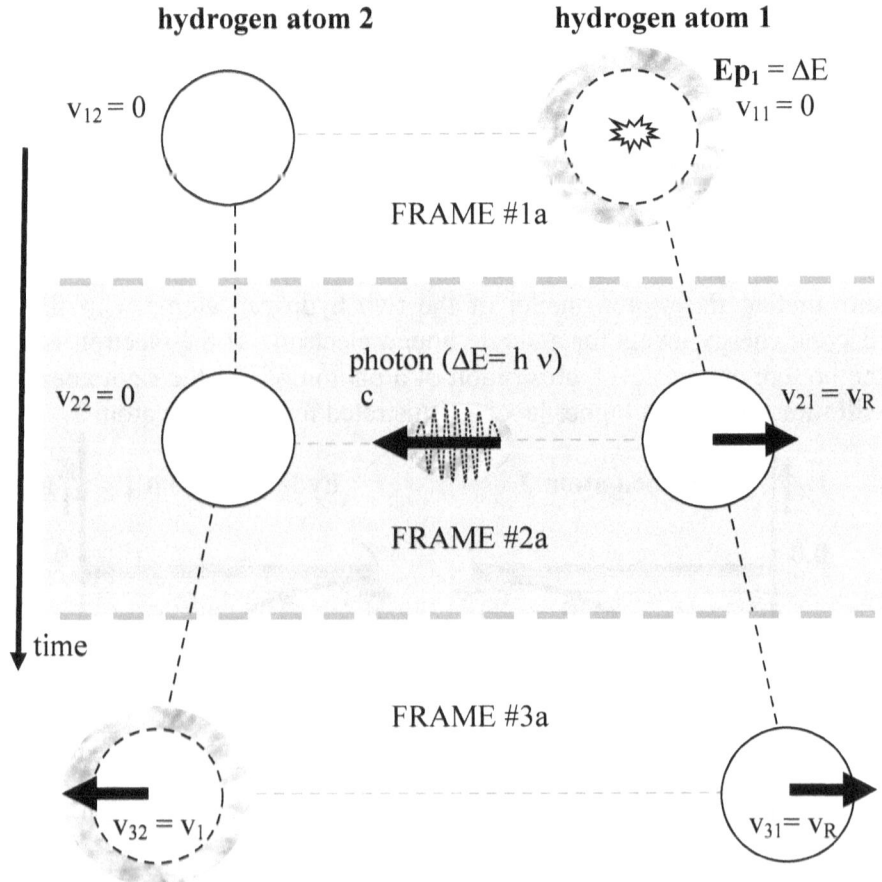

Figure 16.3: Hypothetical dynamics of a non-viable mediated interaction

In this expression h is Planck's constant defined earlier and v is the frequency of the radiation associated with the photon. The relationship of the frequency and wavelength of radiation is given by:

$$v = c / \lambda$$

with λ defined as the wavelength of the radiation, and of course c is the universal speed of light through a vacuum.

By virtue of its higher internal energy state after having absorbed a photon, the rest mass of a molecule that we defined as m_o before absorption will be increased to $m_o + \Delta E$. The resulting more massive molecule should now have a rest mass that we will define as M_o that is greater than m_o by virtue of being in a higher energy state as follows:

$$M_o = m_o + h \, v / c^2$$

So that it possesses the following total energy:

$$E = M_o \, c^2 = m_o \, c^2 + h \, v$$

However, if it were in relative motion v, its effective mass would be:

$$M = M_o / (1 - v^2/c^2)^{\frac{1}{2}} = c^2 (m_o + h \, v / c^2) / (1 - v_o^2/c^2)^{\frac{1}{2}}$$

Applying the conservation equations to the situation depicted in figure 16.3, we obtain the following for each frame:

FRAME #1a:

$$E_{TOTAL1} = m_o \, c^2 + (m_o \, c^2 + h \, v)$$

$$p_{TOTAL1} = 0$$

FRAME #2a   (conceptualized only)

$$E_{TOTAL2} = m_o \, c^2 + h \, v + m_o \, c^2 / (1 - v_R^2/c^2)^{\frac{1}{2}}$$

$$p_{TOTAL2} = - h \, v / c + m_o \, v_R / (1 - v_R^2/c^2)^{\frac{1}{2}}$$

FRAME #3a:

$$E_{TOTAL3} = m_o \, c^2 / (1 - v_R^2/c^2)^{\frac{1}{2}} + (m_o \, c^2 + h \, v) / (1 - v_1^2/c^2)^{\frac{1}{2}}$$

$$p_{TOTAL3} = - (m_o + h \, v/c^2) \, v_1 / (1 - v_1^2/c^2)^{\frac{1}{2}} + m_o \, v_R / (1 - v_R^2/c^2)^{\frac{1}{2}}$$

Using exclusively the conservation of momentum equations above and equating total momentum, we obtain for FRAMES #1a and #2a:

$$h \, v / c = m_0 \, v_R / (1 - v_R^2/c^2)^{\frac{1}{2}}$$

So for this initially relatively-stationary case, we obtain for v_R:

$$v_R = H / (1 + H^2/c^2)^{\frac{1}{2}}$$

where we have defined:

$$H = h \, v / m_0 c \geq v_R$$

And since, $m_0 \, v_R / (1 - v_R^2/c^2)^{\frac{1}{2}} = h \, v / c$,

momentum considerations from frames #2 and #3, should result in:

$$h \, v / c = (m_0 + h \, v/c^2) \, v_1 / (1 - v_1^2/c^2)^{\frac{1}{2}}$$

$$H - (1 + H/c) \, v_1 / (1 - v_1^2/c^2)^{\frac{1}{2}}$$

$$v_1^2 = [H^2 / (1 + H/c)^2] (1 - v_1^2/c^2)$$

$$v_1^2 (1 + [H^2 / (1 + H/c)^2] / c^2) = H^2 / (1 + H/c)^2$$

So that for v_1 we would obtain:

$$v_1 = I / (1 + I^2 / c^2)^{\frac{1}{2}} < v_R$$

where we define:

$$I = H / (1+H/c) \geq v_1$$

But that is not the complete (nor even a consistent) picture. The problem to be solved, as was the case regard to Boltzmann's analyses, is the determination of values for the transfer of energy and momentum from one molecule to the other. In this case, however, knowing the velocities v_{11} and v_{12} we must determine v_{31} and v_{32} of the atoms #1 and #2 in the first and final frame to assure that the initial binding energy possessed by atom #1 as well as a portion of its initial kinetic energy (if any) have been transferred to atom #2. Refer to FRAMES #1a and #2a here as well as similar frames in figures 16.5 and 16.6 as merely conceptualized first steps in obtaining the resulting velocities. This procedure must also include consideration of involved momentum exchanges. Then we must do the same thing for FRAMES #2a and #3a to resolve the same issues with regard to the absorption of the photon by atom #2. Equations pertain here to an originally stationary reference frame, i. e., they would have to be expressed differently in the initial reference frame of atom #1 than they would in the eventual frame, etc..

But all these are moot points; the conservation of energy cannot be made to apply for zero initial relative velocity so the interaction cannot occur. It is an inviable interaction. So naturally there is also no possible reverse process either.

conservation difficulties with the mediation process

Quite clearly molecular interactions are more diverse and more complicated than was specified by Boltzmann or later by Einstein in his quantum theory of radiation. Certainly it is more involved than elastic collisions of material objects. Clearly also there is much more involved than either the 'stimulated' or 'spontaneous' emission discovered in regard to Einstein's quantum theory of radiation.

In the following more specific example we address the possibility of emission of a photon from a 'stationary' molecule. This would be analogous to Einstein's 'spontaneous' emission. What we will find is that intermediate accounting of energy and momentum is fraught with inevitable problems. We will attempt to address momentum and energy considerations for the emission of a photon from an isolated molecule similar to the situation in figure 16.1 to show that it is impossible.

In this scenario, which we will illustrate in figure 16.4, we have very similar energy and momentum conservation considerations to the situation of baseball players. However, when the application of conservation laws involves photon emission there are the additional issues to be addressed. Not least of these is the fact that photons and material particles exhibit a different relationship between the associated energy and momentum expressions as discussed previously. This difference complicates the analysis of interactions between particulate matter and radiation. Consider, for example, the simple case of an atom or molecule emitting a photon as shown in figure 16.4 below. Here we have shown the molecule or atom that is in its higher energy state as possessing an enlarged cloud about a central core. This is just an artifact of the illustration that will help us distinguish which molecule is the prospective emitter and which the absorber.

In a frame of reference of the atom or molecule that is emitting the photon we have:

$$E_1 = m_o c^2 + \Delta E$$
$$p_1 = 0$$
$$\Delta E = h\nu$$

$$E_1 = E_{2p} + E_{2m}$$

FRAME #1

$$p_1 = p_{2p} + p_{2m}$$

FRAME #2

So, since $\Delta E = h\nu$,

$$h\nu = (m_o c^2 + h\nu) - m_o c^2 / (1 - v_R^2/c^2)^{1/2}$$

and

$$h\nu/c = m_o v_R /(1 - v_R^2/c^2)^{1/2}$$

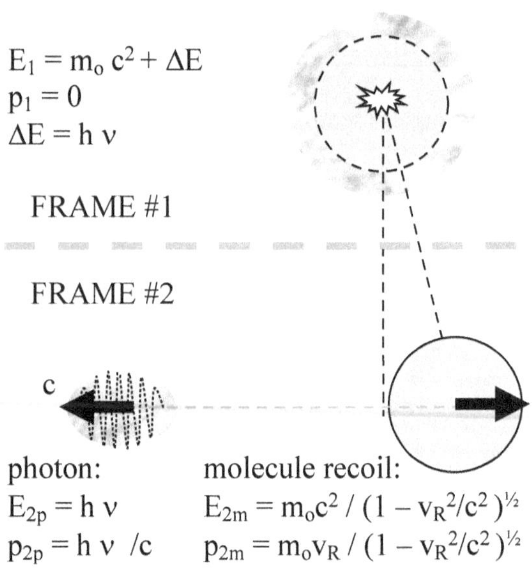

photon:
$$E_{2p} = h\nu$$
$$p_{2p} = h\nu /c$$

molecule recoil:
$$E_{2m} = m_o c^2 / (1 - v_R^2/c^2)^{1/2}$$
$$p_{2m} = m_o v_R / (1 - v_R^2/c^2)^{1/2}$$

Figure 16.4: Photon-particle interaction concept

Because $E_2 > E_1$, if we were to try to solve these equations for the recoil velocity v_R, we would obtain from this sequence of expressions:

$$v_R/c = 1$$

Thus we are led ultimately to an untenable conclusion that $v_R = c$. It was suggested earlier with regard to figure 14.3 that showed that relationships between momentum and energy expressions for particles are totally incompatible with that of photons except in the most extreme and unrealistic case for which $v = c$. However, there is no such extreme case here since the frequency of the radiation given off by atoms can be very low so that associated recoil velocities could in fact be near zero.

As is the case with science generally, we must doubt our analyses and attempt to refute our own arguments. We have done that and found that it is not the analysis that is at fault here. The problem, as suggested in chapter 14, is much more basic than that. There is no way for a photon (a quasi-particle) to be separately emitted from or absorbed by relatively stationary atoms and have both energy and momentum conserved throughout the interaction! So what is the solution?

Major physicists have not been totally oblivious to this problem throughout the 20th century. In particular Lewis[10] noted in 1926 that:

---

[10] G. N. Lewis, "The Nature of Light", *Proceedings of N. A. S.*, 12, 22-29 (1926)
Lewis went so far as to say: "I propose to eliminate the idea of mere emission of

"...we can no longer consider one atom the active agent and the other as an accidental and passive recipient, but both atoms must play coordinate and symmetrical parts in the process of an exchange."

And as stated elsewhere by Bonn[11] in citing the work of Wheeler and Feynman[12,13], and Cramer[14,15]:

"Lewis's position was notably cited by Wheeler and Feynman in their analyses of light as an inter-particle interaction in contrast to its being just another object or 'wave/particle duality'.
"... Cramer did ... address this misconception in his Transaction Interpretation of Quantum Mechanics."

further details of the emission/absorption process

So clearly part of the fault with the analysis with regard to figure 16.4 above is the attempt to treat emission of a photon as a separated phenomenon for analysis. Photon interactions cannot be treated in that way. Emission and absorption phenomena are inextricably linked.

Clearly there is more to this problem than meets the eye and more subtlety than covered in Einstein's brief treatment of the quantum theory of radiation. In quantum theory there is a continuous wave function aspect that is similar to the classical picture illustrated earlier as figure 14.1 but there is also a *wave function collapse* aspect that occurs when one of a multitude of possible absorbers of the radiation actually *does* absorb it. When that occurs, and only then, does the spherical wave aspect totally disappear. These are two aspects of the notable 'wave/particle' duality of radiation.

Wheeler and Feynman's absorption theory exploits both a 'retarded' as well as the 'advanced' electromagnetic wave (and to similar effect with the bidirectional wave function in quantum theory). These two wave functions allowed by Maxwell's equations and also

---

light and substitute the idea of ***transmission***, or a process of exchange of energy between two definite atoms or molecules."

[11] R. F. Bonn, ***The Aberrations of Relativity***, Vaughan Pub., Seattle, p. 14 (2008)

[12] J. A. Wheeler and R. P. Feynman, "Interaction with the Absorber as the Mechanism of Radiation," *Rev. Mod. Phy.*, 17, 157 (1945)

[13] J. A. Wheeler and R. P. Feynman, "Classical Electrodynamics in Terms of Direct Interparticle Action," *Rev. Mod. Phy.*, 21, 425 (1949)

[14] J. G. Cramer, "The transaction Interpretation of Quantum Mechanics," ***Rev. Mod. Phys.***, 58,3, 647-687 (1986);

[15] J. G. Cramer, "Generalized Absorber Theory and the Einstein-Podolsky-Rosen Paradox," ***Phys. Rev. D***, 22, 2, 362-376 (1980).

current quantum theory can be seen as proceeding forward and backward in time respectively. If we accept that notion, then...

"It works like this. When an electron [within a molecule in our case] *vibrates, on this picture, it attempts to radiate by producing a field which is a time-symmetric mixture of a retarded wave propagating into the future and an advanced wave propagating into the past. As a first step in getting a picture of what happens, ignore the advanced wave and follow the story of the retarded wave. This heads off into the future until it encounters an electron* [again within a molecule in our case] *which can absorb the energy being carried by the field. The process of absorption involves making the electron that is doing the absorbing vibrate, and this vibration produces a new retarded field which exactly cancels out the first retarded field. So in the future of the absorber, the net effect is that there is no retarded field."* [16]

Well... okay, so that's pretty obscure. See figure 16.5 for a graphic depiction. What is very clear however is that it takes two to tango in a mediated interaction. Without a receptive molecule to absorb radiation emission is incomplete; further, as we will see, unless that receptive molecule is approaching, relative the emitting molecule, no interaction can occur at all.

The significant conclusion is that there is no net disassociation of a photon from the emitting molecule unless and until that photon of radiation is actually committed to a specific interaction. Further, it should therefore be concluded that as Einstein posited in his Special Theory of relativity, it is indeed from the perspective of the observer, i. e., the molecule that will actually absorb the photon in this case, that we must analyze this process. The transfer of energy and momentum from one molecule to another is not completed and cannot be accounted properly using conservation laws until the process is totally completed.

So the very concept of relativity is somewhat at issue here. It seems that the 'observer's' role is of primary concern to the analyses that must be performed. This one-sided perspective means that relative velocity must always be treated as the velocity of the emitter relative to the absorbing molecule rather than the other way around. Otherwise we get into the problem identified in figure 16.4 with a relatively stationary emitter. We will see this play out where the conservation laws can be satisfied only if the absorber's perspective is assumed throughout.

---

[16] John Gribbin, *Q is for Quantum – An Encyclopedia of Particle Physics*, The Free Press, New York, P.411 (1999)

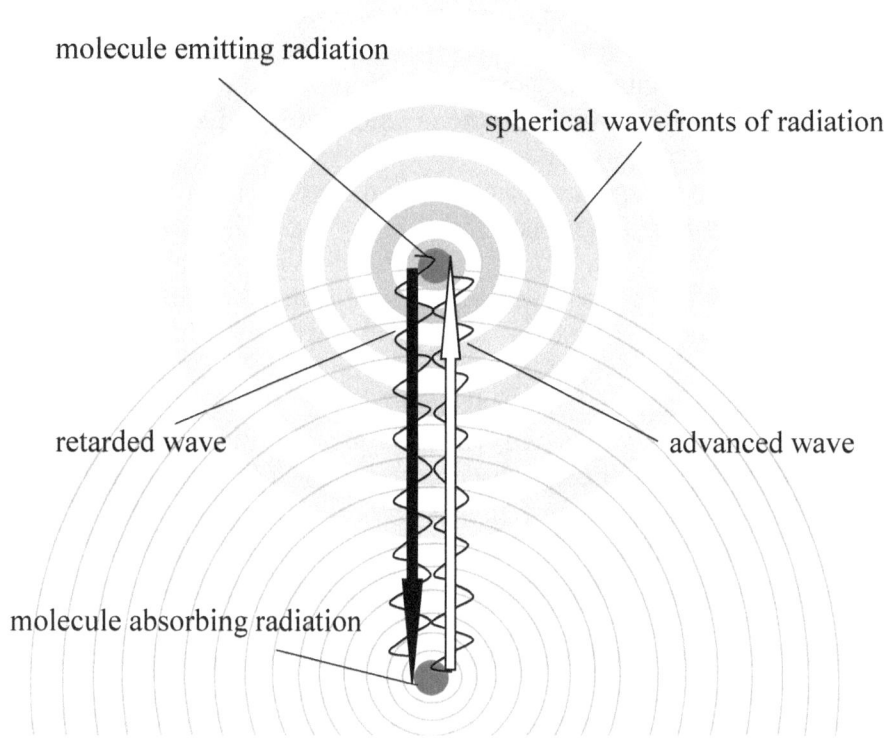

Figure 16.5: Retarded and advanced wave functions of emitter and absorber

Photons are obviously not objects to be caught in mid-flight. They are but part of a total transaction for which conservation laws must apply *upon completion* – if (and only if) the interaction is viable. Any conception of intermediate steps in this process will be fraught with inevitable inaccuracies. Nonetheless we proceed to conceptualize interactions as three-step processes to provide a vague notion of what is transpiring during the transaction. We can apply conservation of momentum throughout the steps but the conservation of energy in particular will remain problematic as we saw above unless we have a completely viable transaction.

the issue of 'spontaneous' and 'stimulated' emission

As noted in the previous chapter, Einstein's efforts to derive the blackbody radiation distribution based on his quantum theory of radiation relied on there being two distinct types of radiation emission. He addressed the probabilities of the two types from which his conclusion with regard to thermal radiation was secured. But he made no attempt to envision any mechanism for how these two categories could come about. We take a certain amount of exception to the notion of there being two types, not just on the logical grounds stated earlier with regard to embracing two distinct processes for emission of identical radiation, but from a more practical perspective of how would the two types actually occur.

With regard to 'spontaneous' emission we have noted above, based on the conservation laws, that it is virtually impossible to envision an emission to take place without there being an associated absorber. The conservation laws cannot be satisfied by an isolated molecule emitting a photon without regard to what will eventually become of it. Furthermore, it is the absorber that accommodates any such transaction by having a compatible dynamical situation.

The concept of 'stimulated' emission involves molecules that exist in a state of having already absorbed a photon but are stimulated by other radiation of the same frequency, causing them to emit photons. Thus a situation arises with two emitted photons where there had been one, etc.. Thus monochromatic radiation can be amplified as has been exploited by laser and maser technologies. In thermodynamics, however, it is merely the amount of heat energy in the system that expands the number of molecules in their high state and with it the probability that an emitted photon will encounter a molecule in its high state rather than one in a low (receptive) state with which to resonate.

viability of 'spontaneous' emission with relative motion

The discussion so far of two atoms or molecules at rest with respect to each other is not a typical situation. In any thermodynamic system at a temperature above absolute zero there are predominantly constituents at the submicroscopic level that are in relative motion.

So let us consider the situation depicted in figure 16.6 in which the two molecules approach each other at the relative velocity of v_R. In FRAME #2b of this figure the molecule #1 has already emitted a photon and has recoiled in accordance with the conservation of momentum such that the two molecules are *now* stationary with respect to each other with photon transmission underway. We selected a value for v_R to guarantee this result. It is a totally unique situation that we will find to be a limiting case – and an unusual exception in as much as it actually accommodates velocity reversibility.

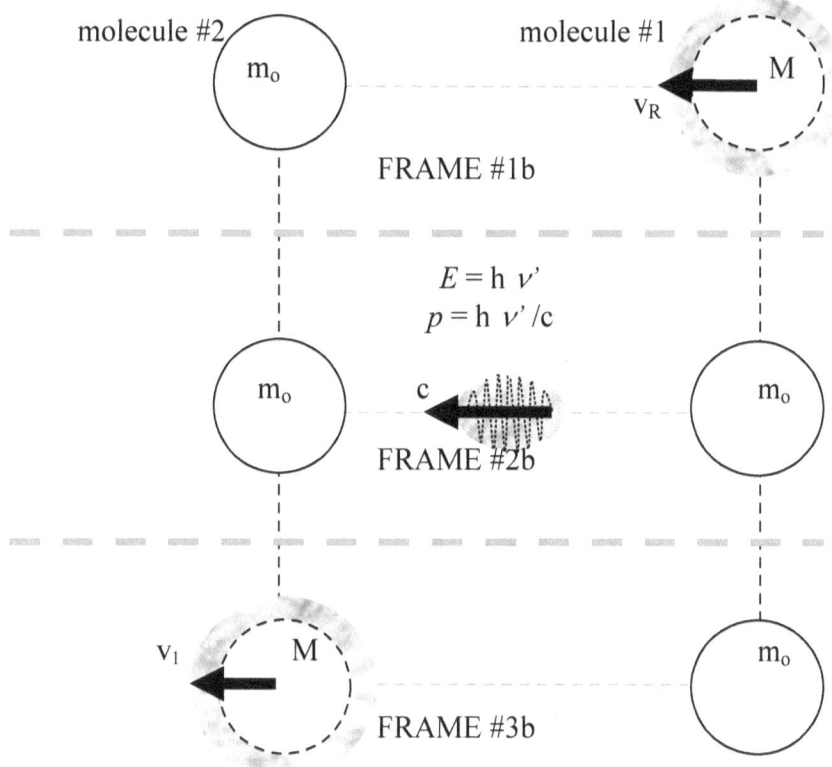

Figure 16.6: 'Spontaneous' emission with line-of-sight relative motion

However, the frequency (tantamount to the energy) of the emitted radiation will have increased because of the initial velocity with regard to the eventual absorber, molecule #2, because of the Doppler Effect. Since they were in relative motion at the moment the photon was emitted the effect must be included in the analyses.

Because the molecules are at rest *after* emission it might seem as though the radiation should not be affected by the Doppler Effect, but because of the initial relative motion, it will in fact be Doppler shifted for molecule #2. The 'observer' in Einstein's special relativity must be associated with the absorber; the frequency of the absorbed radiation *will* be increased by the Doppler Effect due to the approach velocity of the original source with respect to the absorbing molecule. Thus, for this case of relative motion of the emitter and absorber the wavelength λ is decreased and the frequency ν is increased. In classical Newtonian mechanics, this change is in accordance with the relationship:

$$\nu' = \nu \left(1 - v_R/c \cos \theta \right)$$

Here θ is the angle to the source of the emission with respect to the direction of relative motion. But the velocity v_R is negative in this case.

The mass of an emitting molecule will be reduced by the equivalent mass energy of the photon just emitted and the frequency of the photon will be increased with respect to molecule #2. Once the photon is absorbed by the other molecule as shown in FRAME #3b, in addition to the transfer of momentum, the mass of molecule #2 will be increased by the amount of the received photon's equivalent mass energy.

The classical Doppler formula does not rigorously apply in this case even though Einstein himself chose to use it rather than his own relativistic Doppler Effect in his otherwise masterful treatment of the causes of the blackbody radiation distribution that we discussed in the previous chapter and will discuss further. His rationale, we can be sure, was that the difference is a second order effect, involving v^2/c^2, and is, therefore, exceedingly small in virtually all cases of interest. However, we proceed directly to use all applicable conservation laws of his relativity theory that includes, as we already have, the increase in rest mass energy as well as momentum considerations.

In Einstein's Special Relativity the Doppler frequency/wavelength expression becomes the following:

$$v' / v = (1 - v_R^2/c^2)^{1/2} / (1 - v_R/c \cos \theta)$$

Thus, for the ratio of the wavelengths $\lambda = c/v$ and $\lambda' = c/v'$ of the same photon from the two perspectives is:

$$\lambda' / \lambda = (1 - v_R/c \cos \theta) / (1 - v_R^2/c^2)^{1/2}$$

So that for a directly approaching source of a photon of radiation:

$$v' / v = (1 - v_R^2/c^2)^{1/2} / (1 - v_R/c) = \sqrt{ (1 + v_R/c) / (1 - v_R/c) }$$

The appropriate expression above for a directly receding source of the photon of radiation becomes:

$$v' / v = (1 - v_R^2/c^2)^{1/2} / (1 + v_R/c) = \sqrt{ (1 - v_R/c) / (1 + v_R/c) }$$

Since we selected the initial velocity v_R of the molecule #1 to force its initial *kinetic* energy to correspond with the eventual total energy of the photon after emission, molecule #1 will then be left with zero velocity having transferred its momentum to the emitted photon. Thus, following the emission of the photon, both molecules will be stationary with respect to each other; all of the kinetic energy from molecule #1 as well as its momentum will have been transferred to the propagating photon.

FRAME #1b:

$$E_{TOTAL1} = m_o c^2 + (m_o c^2 + h \nu) / (1 - v_R^2/c^2)^{\frac{1}{2}}$$

$$p_{TOTAL1} = - (m_o + h \nu / c^2) v_R / (1 - v_R^2/c^2)^{\frac{1}{2}}$$

FRAME #2b:

$$E_{TOTAL2} = 2 m_o c^2 + h \nu (1 - v_R^2/c^2)^{\frac{1}{2}} / (1 + v_R/c)$$

$$p_{TOTAL2} = - h \nu (1 - v_R^2/c^2)^{\frac{1}{2}} / c (1 - v_R/c)$$

FRAME #3b:

$$E_{TOTAL3} = m_o c^2 + (m_o c^2 + h \nu) / (1 - v_1^2/c^2)^{\frac{1}{2}}$$

$$p_{TOTAL3} = - (m_o + h \nu/c^2) v_1 / (1 - v_1^2/c^2)^{\frac{1}{2}}$$

Again, the law of the conservation of energy assures us that the total energy E_{TOTAL} of all components before and after each frame described graphically in figure 16.6 must remain the same for any viable interaction. Likewise, the law of the conservation of momentum assures us that the total momentum p_{TOTAL} of all components before and after each frame described graphically above must also remain the same. However, additional energy $h\nu'$, greater than $h\nu$, will be taken up as kinetic energy by molecule #2 since internally it can only accept $h\nu$. Significantly, we never actually have to use equations of FRAME #2 which remain somewhat suspect. But we will use the momentum conservation equations from FRAMES #1b and #2b as follows:

$$(m_o + h \nu / c^2) v_R / (1 - v_R^2/c^2)^{\frac{1}{2}} = h \nu (1 - v_R^2/c^2)^{\frac{1}{2}} / c (1 - v_R/c)$$

Such that:

$$h \nu (1 + v_R/c) / c = (m_o + h \nu/c^2) v_R$$

Then directly equating energy and momentum of FRAMES #1b and #3b, we ultimately obtain:

$$v_1 = v_R = H = h \nu / m_o c,$$

As we might have expected, the initial and final velocities are simply exchanged by the molecules. It is this complete exchange that is similar to Boltzmann's elastic collisions that assures reversibility.

Of course this recoil velocity parameter is extremely small by comparison with velocities of molecules in a thermal gas in virtually every case since Planck's constant is extremely small, $h = 6.626$ x 10^{-27} erg sec and of course atomic/molecular masses are small as well. Certainly the hydrogen atom is the smallest at $m_o = 1.67$ x 10^{-24} gram. So that:

$$H = 1.323 \text{ x } 10^{-14} \text{ } v = 3.98 \text{ x } 10^{-4} / \lambda$$

Thus even for the Lyman-alpha photon emission line for which the wavelength is $\lambda = 1.216$ x 10^{-5} cm, we would still have the small velocities of $v_1 = v_R \approx 32.7$ cm/sec. This is to be contrasted with the most probable speed of a molecule in a thermal gas that is determined by the temperature of the gas in accordance with the formula we derived earlier of $S_{mp} = (2 \text{ } k \text{ } T / m_o)^{1/2} = 1.2856$ x $10^4 \text{ } T^{1/2}$, where $k = 1.38$ x 10^{-16} erg/Kelvin is Boltzmann's constant. So even at room temperatures for which $S_{mp} > 2$ x 10^5, we have $v_R \approx 10^{-4} S_{mp}$.

viability of 'stimulated' emission with relative motion

The case of 'stimulated' emission cannot be that different from the situations involving 'spontaneous' emission since it possesses after all the very same radiation characteristics in both cases. It must be in the purview of the receiving (absorbing) molecule that distinguishes the process if they are to be distinguished. In this case, however, a molecule that already is in the high state becomes involved as a quasi-absorber, albeit with additional implications caused by resonation which is more directly reflected in the depiction of 16.5 that any other since the photon aspect of the quantum transmission process has not yet been fully determined..

But let us consider the situation as depicted in figure 16.7 in which two molecules had approached each other at the relative velocity of v_R. In FRAME #1' of this figure the molecule #1 has already emitted a photon and has recoiled however might have been appropriate in accordance with the conservation of momentum such that the two molecules are *now* stationary with respect to each other with photon transmission underway. Again we selected a value for v_R to guarantee this result. It is again a completely unique situation that is a limiting case that again accommodates velocity and time reversibility.

molecule #2   molecule #1

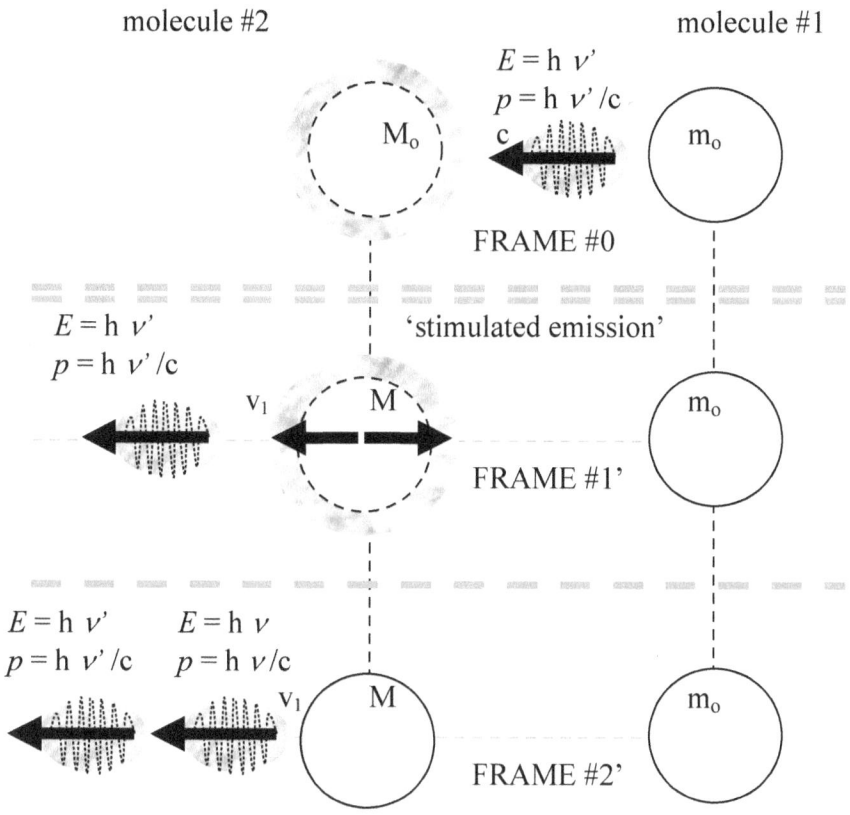

Figure 16.7: 'Stimulated' emission with line-of-sight relative motion

However, the frequency (tantamount to the energy) of the emitted radiation will have increased because of the initial velocity with regard to the eventual 'absorber', molecule #2, because of the Doppler Effect. We will discuss resonance later in regard to scattering phenomena that we will show to be more directly related to 'stimulated' emission.

somewhat more general case of 'spontaneous' emission

Let us get a little more general with regard to 'spontaneous emission and consider a case in which one of the molecules possesses a relative velocity that is not precisely equal to that which could be nullified by the reaction to emitting a single photon but retaining the restriction that $\theta = 0$ nonetheless, so that relative motion of the molecules is still approaching along the direction of their centerlines. This situation is illustrated in figure 16.8. We will proceed frame by frame applying the conservation laws of transfer of energy and momentum from one atom/molecule to the other, again in much the same way we demonstrated in figure 16.1.

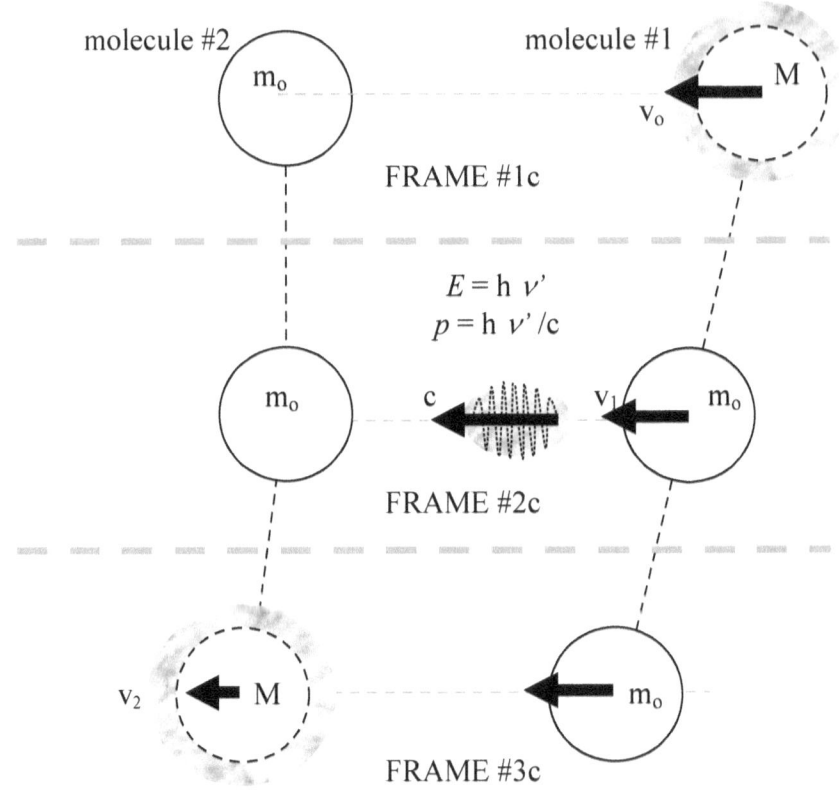

Figure 16.8: Emission-absorption transaction with line-of-sight relative motion

FRAME #1c:

$$E_{TOTAL1} = m_o c^2 + (m_o c^2 + h\nu)/(1 - v_o^2/c^2)^{1/2}$$

$$p_{TOTAL1} = -(m_o + h\nu/c^2)v_o/(1 - v_o^2/c^2)^{1/2}$$

FRAME #2c:

$$E_{TOTAL2} = m_o c^2(1 + 1/(1 - v_1^2/c^2)^{1/2}) + h\nu(1 - v_o^2/c^2)^{1/2}/(1 + v_o/c)$$

$$p_{TOTAL2} = -h\nu(1 - v_o^2/c^2)^{1/2}/c(1 + v_o/c) - m_o v_1/(1 - v_1^2/c^2)^{1/2}$$

FRAME #3c:

$$E_{TOTAL3} = m_o c^2/(1 - v_1^2/c^2)^{1/2} + (m_o c^2 + h\nu)/(1 - v_2^2/c^2)^{1/2}$$

$$p_{TOTAL3} = -m_o v_1/(1 - v_1^2/c^2)^{1/2} - (m_o + h\nu/c^2)v_2/(1 - v_2^2/c^2)^{1/2}$$

In FRAME #2c molecule #1 has already emitted a photon and recoiled in accordance with the conservation of momentum such that the two molecules now experience a different relative velocity v_1 with respect to each other rather than v_0 with photon transmission underway. In FRAME #3c the process is completed and we can equate the energy and momentum of FRAME #1c and FRAME #3c. Equating first the energy and simplifying, we obtain,

$$\gamma_1 = (\gamma_0 - \gamma_2)(1 + H / c) - 1$$

Here we are associating the γ_i symbols with the three relative velocities, v_0, v_1, and v_2 as follows:

$$\gamma_i = 1 / (1 - v_i^2/c^2)^{\frac{1}{2}} \text{ and}$$

$$v_i = c (1 + \gamma_i^{-2})^{\frac{1}{2}}$$

Then by simplifying the conservation of momentum from FRAME #1c to FRAME #3c produces:

$$(v_0 \gamma_0 - v_2 \gamma_2)(1 + H / c) = v_1 \gamma_1 = c (1 + \gamma_1^2)^{\frac{1}{2}}$$

Substituting for γ_1 from above, we do indeed obtain an expression for determining v_2 in terms of v_0 and conservation laws are satisfied. So viable solutions to interaction problems do exist in cases such as this.

reversibility of at least one viable centerline interaction

However, we have still dealt exclusively with contrived examples for which $\theta = 0$ which are hardly typical of situations encountered. So far we have looked at only direct centerline 'collision' paths involving atoms or molecules that are in every case approaching each other. There are, of course, very special cases for which unique features are to be expected. Significantly, (other than needing initial kinetic energy to more or less 'prime the pump' or avoid the requirement for a 'free lunch') the situations we have looked at are examples where we expect reversible submicroscopic interactions. As a particular example, let us look at reversing the situation depicted in figure 16.6. In the initial FRAME of figure 16.9 we illustrate the velocity reversal of FRAME #3b of figure 16.6. The same goes for Frame #2d reversing FRAME #2b and also for Frame #3d, which of course reverses FRAME #1b. And essentially the result is the same as for the situation depicted in figure 16.6 except that now everything is proceeding in the opposite direction.

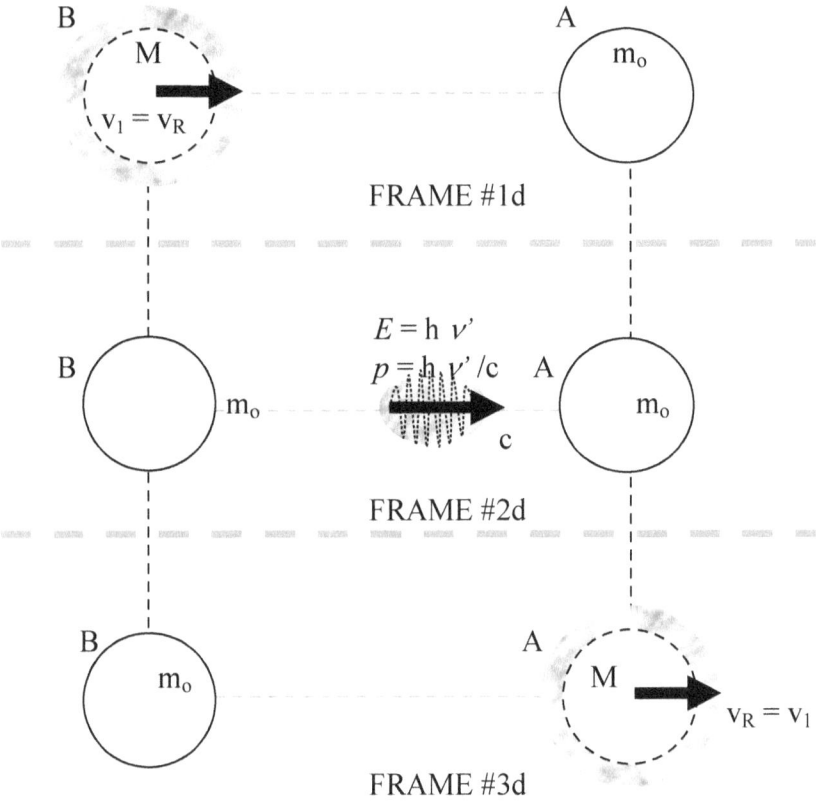

Figure 16.9: Reversibility of viable emission-absorption transaction

FRAME #1c:

$$E_{TOTAL3} = m_o c^2 + (m_o c^2 + h \nu) / (1 - v_1^2/c^2)^{\frac{1}{2}}$$

$$p_{TOTAL3} = M v_1 = (m_o + h \nu/c^2) v_1 / (1 - v_1^2/c^2)^{\frac{1}{2}}$$

FRAME #2c:

$$E_{TOTAL2} = 2 m_o c^2 + h \nu (1 - v_R^2/c^2)^{\frac{1}{2}} / (1 + v_R/c)$$

$$p_{TOTAL2} = h \nu (1 - v_R^2/c^2)^{\frac{1}{2}} / c (1 + v_R/c)$$

FRAME #3c:

$$E_{TOTAL1} = m_o c^2 + (m_o c^2 + h \nu) / (1 - v_R^2/c^2)^{\frac{1}{2}}$$

$$p_{TOTAL1} = M v_o = (m_o + h \nu/ c^2) v_R / (1 - v_R^2/c^2)^{\frac{1}{2}}$$

Thus, even with interactions mediated by radiation we have the verifiable claim that at least one viable submicroscopic interaction is also reversible. It might seem to be the same for mediated atomic and molecular interactions generally as it is for the baseball players or elastic collisions of billiard balls and Boltzmann's molecules. But that is definitely not the general case.

conditions of viability, reversibility, and irreversibility

We saw in the discussion pertaining to figures 16.3 and 16.4 that no viable transmission can occur between two molecules if they are relatively stationary or receding from one another prior to emission of a photon from one to the other. The reason for this is that the recoil of emitting a photon results in the emitting molecule and the photon it emits having more total energy than the pre-emission molecule by itself. So although the conservation of momentum equations can be questionably satisfied, the conservation of energy cannot. So such interactions cannot occur.

The pre-absorption (or pre-wave-function-collapse) phase of this process does not exhibit this problem in classical or quantum mechanics. However, to posit that emission can only have occurred upon completion of absorption is somewhat problematic in its own right. Later we will address aspects of this phenomena related to 'stimulated emission'.

Figures 16.6 and 16.9 are examples of a complete one-step transfer of energy to the other molecule; these two complementary interactions also happen to be the velocity and time reversible version of the other interaction in the strictest sense of the term reversibility. But these two reversible interactions are extremely unique situations in which the initial relative velocity and therefore momentum of one molecule relative to the other precisely equals the recoil momentum of the emitted photon. Of course following emission of the photon, there will be no net relative velocity, but this is a very transitory situation that pertains only until absorption has taken place.

At least when there are appreciable velocities of approach we will find viability of interactions and these are interactions that can at least be repeated in the sense of it being possible for a photon to be returned from the absorber back to the original emitter as long as the molecules are still approaching. Increasingly, however, more of the energy will be transferred between molecules as the result of such repeated exchanges of photons until, as described in the previous paragraph, all of the energy is transferred to the other molecule. But this does not constitute reversibility except for a final transfer as described in the previous section of this chapter. No time or velocity reversal has been or can be demonstrated for these other cases. Also note that we have addressed the interaction with the emitting molecule assumed to be the one that is in motion.

As neutral atoms or molecules in relative motion approach closer and closer to each other the collisions might be mediated by more photon exchanges prior to an impact that would ultimately reverse the relative velocity so that the molecules would finally be receding from each other rather than approaching. At this point no more photon exchanges can occur. But these exchanges that might precede (or even preclude) collision are not reversible in the sense in which the term is generally applied.

In all the viable interactions we have dealt with so far, the relative velocities of the principals and directions of photon transmissions are approaching along the centerline of the participants. Furthermore, and much more significantly, we have largely ignored the impact of quantum restrictions on frequencies of interchanged photons which totally preclude viability of interactions of molecules that are receding from each other and severely constrain viable angles and velocities of approach. We will address the further restrictions in the next chapter.

17: Inherent Irreversibility of Photon Exchanges

"Although the implications [of the photoelectric effect] *were not fully understood (not even by Einstein) this was the first step towards the concept of **wave-particle duality**. It was, incidentally, for this work, not for either of his theories of relativity, that Einstein received the Nobel Prize.*[17]

Contrary to the predominant misconception, reversible interactions at the submicroscopic level that we have addressed so far are far from typical of submicroscopic interactions that occur in nature. We have discussed photon interactions without regard to all the implications of the major contributions to 'modern physics' made during the first decade of the 20th century by Albert Einstein. What happens when all of the repudiated aspects of classical physics are eliminated?

Einstein's contribution to the Doppler Effect

Of course the awareness of the Doppler Effect preceded Einstein's Special Theory of Relativity by centuries. The change in the pitch of the sound of an approaching or receding train whistle was well known of course, with a similar effect expected and detected for light. That expectation of a change in frequency was as follows:

$$\nu' = \nu / (1 - v/c \, \cos \theta)$$

where parameters are defined as they have been in the preceding chapters with the angle being that of the relative velocity with regard to

---

[17] John Gribbin, *ibi,* P.281 (1999)

the line joining the centerlines as shown in figure 17.1. In the illustration the velocity of approach is negative. Of course radial and transverse velocity components change as functions of the angle θ as the source of radiation approaches or recedes as is shown in figure 17.2.

The rationale for this expectation of classical physics was that the velocity of light was thought to be dependent on the relative velocity of the source and observer of a wave function. With the increased or decreased time of transmission, there would be respectively longer or shorted wavelengths to span the distance from the source to the observer. This expectation was realized experimentally, with the denominator in the previous equation characterizing the anticipated effect of relative motion between an observer and the source for sound vibrations, but that is *not* the case for electromagnetic radiation.

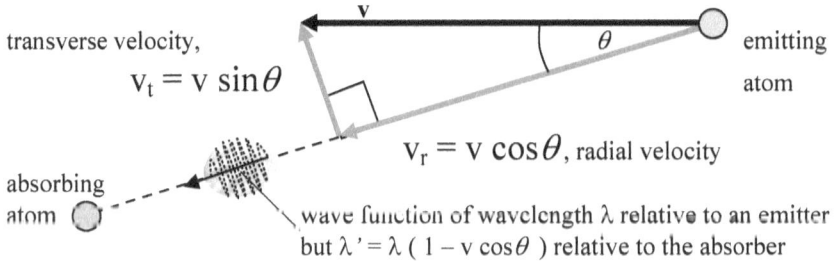

Figure 17.1: Definitions of angle and radial and transverse velocity components

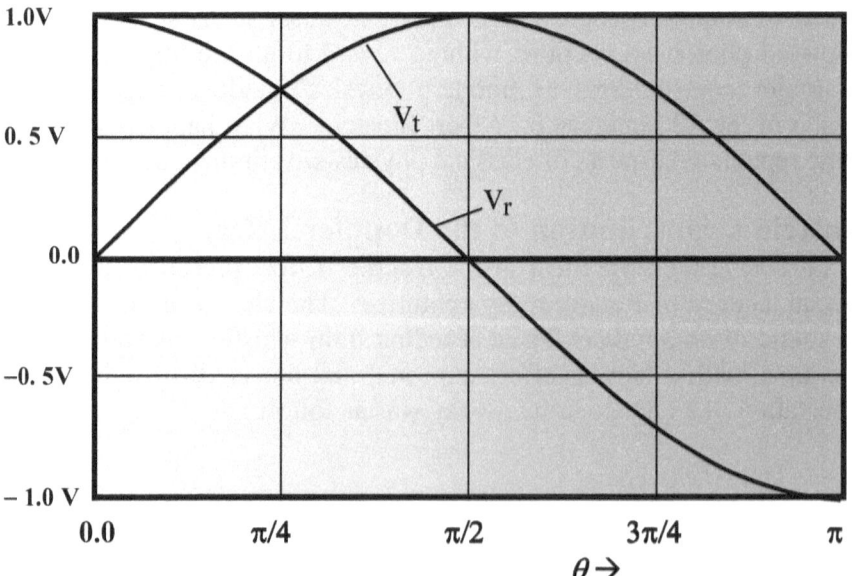

Figure 17.2: Changing values of the transverse and radial velocity components

With the introduction of Einstein's special theory of relativity the expectations for the Doppler Effect on radiation changed. Einstein's *second postulate*, that the velocity of light is the same for all *observers* (and thus for all absorbers of radiation) independent of their relative motion with regard to the source of that radiation in a vacuum, required a different explanation of the observed phenomena. That explanation, like most of what follows in Einstein's theory of relativity can be simply derived from the applicable Lorentz transformation equations. Thus a revised Doppler formula came into being:

$$\lambda' = \lambda\ \gamma\ (\ 1 - v/c\ \cos\ \theta\ )$$

where now there is the additional factor of γ in the formula, with:

$$\gamma = 1\ /\ (\ 1 - v^2/c^2\ )^{1/2}$$

This factor contributes very little to the effect unless the relative velocity v of the source with regard to the observer is significant with respect to the speed of light, c; this is a caveat that is virtually never required on a single interaction. However, when relative velocities are appreciable the effect of the gamma factor can be an observable fact. Refer to figure 17.3 where the magnitude of the Doppler Effect is illustrated for both the classical and relativistic formulas in the case of v/c = ½; in addition the relativistic formula is illustrated for v/c = 0.1, v/c = ¼, as well as for cases where v/c = 0.75, 0.9, 0.99, and 0.999.

Clearly, in the relativistic case when there are interactions between molecules whose relative velocities are *not* aligned with the centerline there is an additional effect to be taken into account. As illustrated in figures 17.3, when $\theta = \pi/2$ there is a frequency change even where none had previously been predicted. Besides a positive or negative recessional Doppler shift in the frequency of a photon, there is a 'transverse' Doppler Effect to be accounted for that was only acknowledged as a reality with the introduction of Einstein's special theory of relativity. Since it is a second order effect in v/c it is therefore extremely small in virtually every case for interacting molecules of gases with temperatures that are typically realized – even in much more extreme cases of high temperature plasma gases in and around the sun and at the centers of large galaxy clusters. However small in any particular instance, the effect modifies the results of virtually *every* interaction and contributes to irreversibility.

However, in addition to the issue of transverse Doppler, there are the ever present quantum effects that also alter the picture.

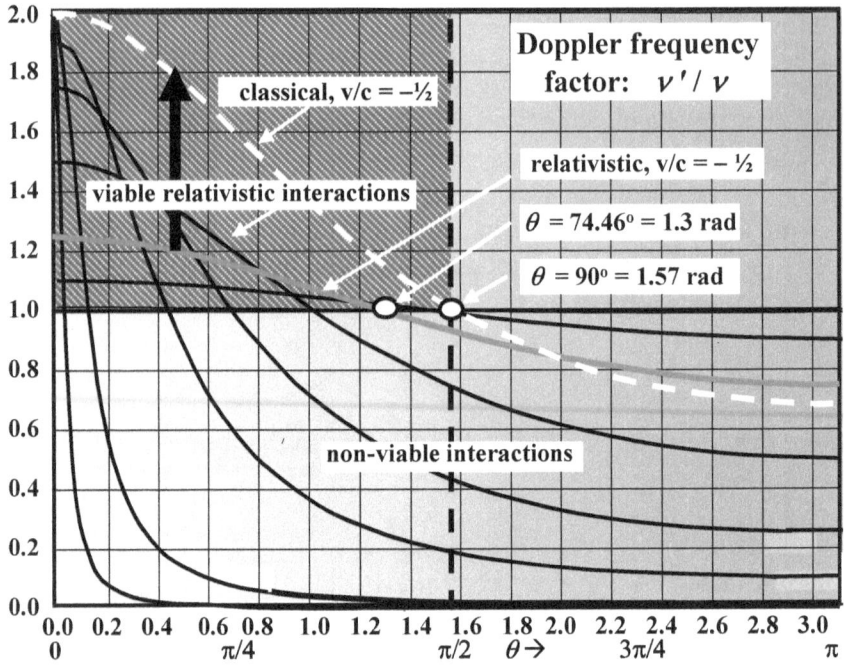

Figure 17.3: Doppler frequency factor in classical and relativistic physics

photoelectric effect constraint on molecular interactions

Among Albert Einstein's major discoveries during the first decade of the 20th century, the 'photo electric effect' is of note here. It was significant to the acceptance that light transmission involves discreet 'chunks' ('quanta') of energy, or 'photons' as G. N. Lewis would denominate them in 1926. The significance of the discovery was that matter can only absorb entire photons of electromagnetic energy.

So let us look at what this means in the context of the Doppler effect: If a photon of a given frequency (energy) is emitted by one molecule, a similar molecule in relative motion can only absorb that photon if the relative velocity is such that the photon has enough energy (in the absorbing molecule's frame of reference) to raise the internal molecular energy to a next quantum level. As we will see, this actually precludes virtually all otherwise allowed interactions from occurring. This is particularly the case where extreme velocities are involved.

The formula for relativistic Doppler frequency v' is the following:

$$v' / v = (1 - v^2/c^2)^{1/2} / (1 - v/c \cos \theta)$$

This formula was plotted in figure 17.3 for a few values of the relative velocity v; the frequency v in the formula corresponds to the energy level difference in a molecule. So if the ratio v'/v is less than unity, then there is insufficient energy for a molecule to be able to absorb such a photon and no interaction results. What this means in essence is that no interaction can occur between molecules that are receding from each other. In fact, for an appreciable relative velocity no interaction will occur unless it is initiated well in advance of the time at which the molecules are at their closest approach. In figure 17.3 that is indicated by the angle at which $v'/v = 1.0$. So the quantum-relativistic criterion for even the possibility of interaction is as follows:

$$0 \geq \cos \theta \geq (1 - (1 - v^2/c^2)^{1/2}) / (v/c)$$

The portions of curves that are compatible with this constraint are shown in the upper left quadrant of the plot in figure 17.3. Solid curves all represent the relativistic formula for approach velocities v=0.999c, v=0.99c, v=0.9c, v=0.75c, v=0.5c, v=0.25c, and v=0.1c, the latter narrowly differs from that for v=0. A direct comparison of the classical physics formula for v=0.5c is illustrated by the dashed curve in the figure. All curves for the classical non-relativistic formula cross the abscissa at $\theta = \pi/2$. In the relativistic case this is only true for v << c.

It is fairly intuitive that molecules that are receding from each other have passed their window of opportunity for interacting. In any case, whether molecules can absorb radiation is determined by quantum theory. This is in combination with relativistic Doppler phenomena. Together they impose rather severe constraints that preclude reversibility in most every case. But we are now equipped to determine combinations of angles, distances, and relative velocities that can at least accommodate molecular interactions. From this vantage we can assess the possibility of viable interactions as illustrated in figure 17.4.

There is also the issue of scattering of radiation whereby transverse velocities of scattering electrons produce redshifting of a continuing photon; in that situation there is a transfer of momentum and energy to the medium. This irreversible phenomenon is in fact the *cause of*, rather than *result of*, the "astrophysical trend" discussed in the forward to this treatise. That subject to the extent that it is responsible for many cosmological effects is discussed exhaustively elsewhere by Bonn.[18]

'classical' physics example of a reversible interaction

Finding example interactions involving transverse motion for which velocity reversal fails to reinstate the former status when the relativistic

---

[18] *Cosmological Effects of Scattering in the Intergalactic Medium*, 2011.

Doppler formula is applied is now straight forward. However, for contrast, let us look at the interaction that would have been expected *if* Einstein's discoveries had not occurred.

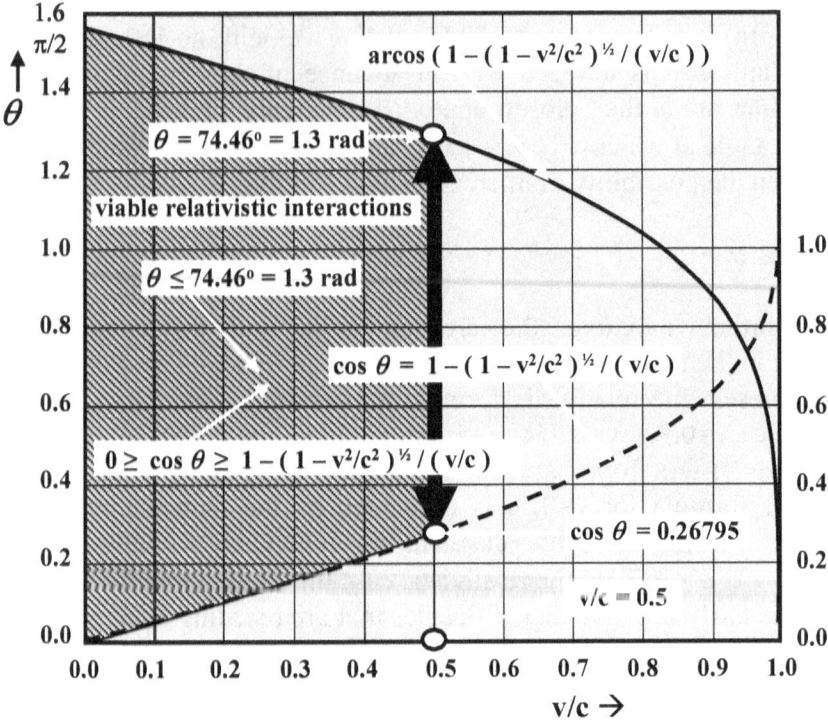

Figure 17.4: Domain of allowed molecular interactions in relativistic physics

Suppose that molecule #1 passes by molecule #2 at a perpendicular distance d. From this position of closest approach it emits a photon of light at an angle that will be perceived as perpendicular to its relative motion such that the light will have traveled the distance d when the two molecules are at their closest as shown in figure 17.6 below.

In the figure at right values of the velocities should be noted. First of all, v_{11} is a given and v_{12} is the vector sum of v_{11} and the recoil velocity perpendicular to it, v_R:

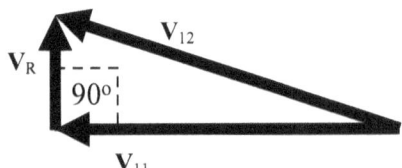

Figure 17.5: Recoil velocity

$$V_{12} = V_{11} + V_R$$

This is as shown in going from frame #1a to frame #2a of panel a in figure 17.6. The photon of light is emitted at a right angle to the path of molecule #1 as allowed by the classical formula plotted for v/c= ½ in figure 17.3. This causes molecule #1 to recoil, leaving its original path. In frame #3a the photon is absorbed by molecule #2 causing it to recoil away from its original location but at an angle of 75°. This recoil is greater due to increased light velocity in a classical analysis.

Doppler Effect in Classical Physics
panel a (forward motion, molecule #2 perspective)

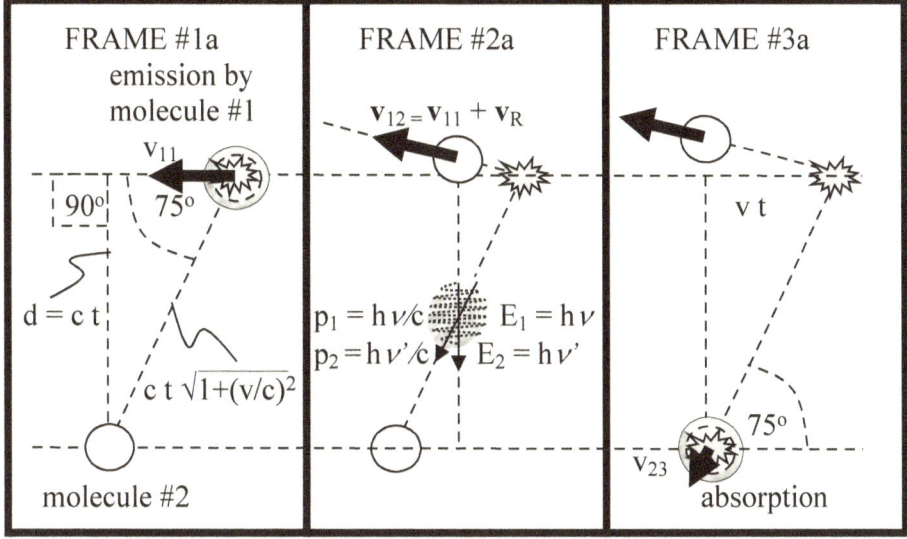

Figure 17.6: Reversible molecular interactions in classical physics

Obviously if relative motion were all that mattered, panel a in figure 17.6 could be drawn from two perspectives: One with molecule #1 in motion, the other with molecule #2 in motion; the angles would be the same in either case. However, as discussed in the previous chapter the analyses must be from molecule #2's perspective. That will be modified further when we apply the aberration of angles in Einstein's relativity. You'll notice that in this rather limiting case where the transverse but approaching velocity becomes a transverse receding velocity similar to the reversible situation in figures 16.5 and 16.7.

Now, if we reverse the direction of all velocities we end up with the situations depicted in figure 17.7. Notice that although light does not actually occupy a specific location in classical analysis (nor for that matter especially in quantum theory) it was known to exhibit momenta in a somewhat similar sense to the current understanding although molecules were not acknowledged as having discreet energy levels at the time. Frame #2b shows restoration of molecule #2 to its 'stationary' role by the recoil of a similar amount to what it received, i.

e., $-v_{23} = v_R / \cos 75°$, which in the frame of molecule #1 will be just the recoil velocity v_R.

Doppler Effect in Classical Physics
panel b (reversed motion, molecule #2 perspective)

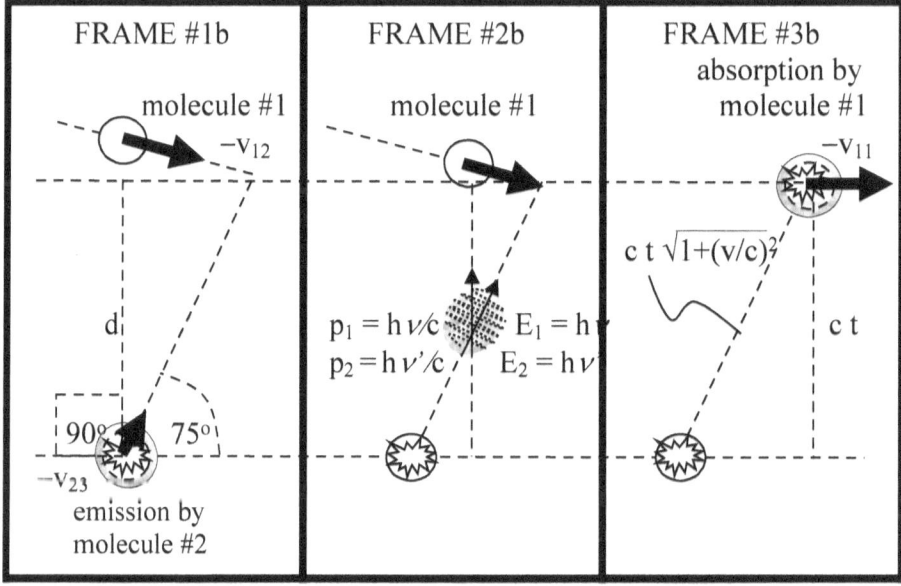

Figure 17.7: **Reversed molecular interaction in classical physics**

In figure 17.8 we illustrate the reversed recoil of molecule #1, where the symmetry between the FRAMEs #1a and #3b, #2a and #2b, and #3a and #1b is complete. All steps in this interaction are completely reversible. It is similar to ballplayers as in figure 16.1.

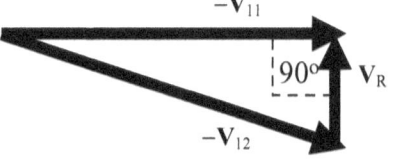

Figure 17.8: **Classical reversal**

In Galilean (classical) relativity the perspectives of the emitter and absorber of radiation, although unique to each, are both completely compatible with the other. This is demonstrated in frames #2a and #2b of figure 17.6 and 17.7 where the photon (if it could be called that) follows two paths between emission and absorption – one appropriate to each perspective. It is as though the 'photon' were a bead sliding along two rods, one connected to each molecule as shown in figure 17.9. The vertical speed is c, the diagonal speed is $c' = c \sqrt{1 + (v/c)^2}$. What is significant here is that both the emission and absorption events, although perceived differently, are in fact the very same in space-time.

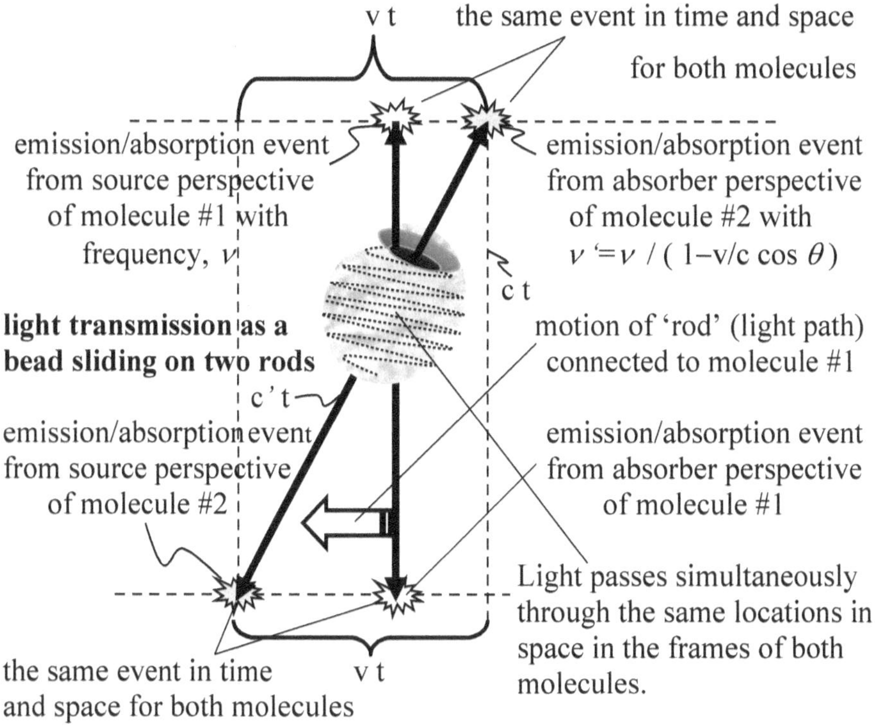

Figure 17.9: Explicit analogy of reversible light transmission in classical physics

relativistic physics necessitates irreversible interactions

In modern physics there are no simplistic mechanical analogies for coordinating emission and absorption events to occur at the same place and time in a single frame of reference. There are no analogous rods to fit in a single hole of anything like an analogous photon bead that would allow it to slide along two perspective directions at once. Respective angles between the emission and absorption events are such that a bead could only be in coincidence at a single point in the entire length of rods whatever the relative velocity as long as it is non-zero. A single such 'bead' could not 'slide' in this way because the two events cannot be simultaneous in relatively moving frames.

Resolution of this disparity in the times and positions of what are conceived to be identical emission and absorption events in Einstein's relativity relies on scale differences in the space-time of relatively moving 'observers'. Rather than a difference in the speed of light for two observers, there is a difference in the amount of time from emission to absorption as well as a difference in position of emission. So, with relativistic Doppler and quantum formulas, both introduced by Einstein, the situation of light transmission illustrated in figures 17.6 through

213

17.9 is invalidated. The two emission events, drawn separately in figure 17.9 but notably coincident cannot be considered simultaneous or coincident in Einstein's relativity; they are separate (although in some conceptual sense, the same) events in space-time as shown in figure 17.10.

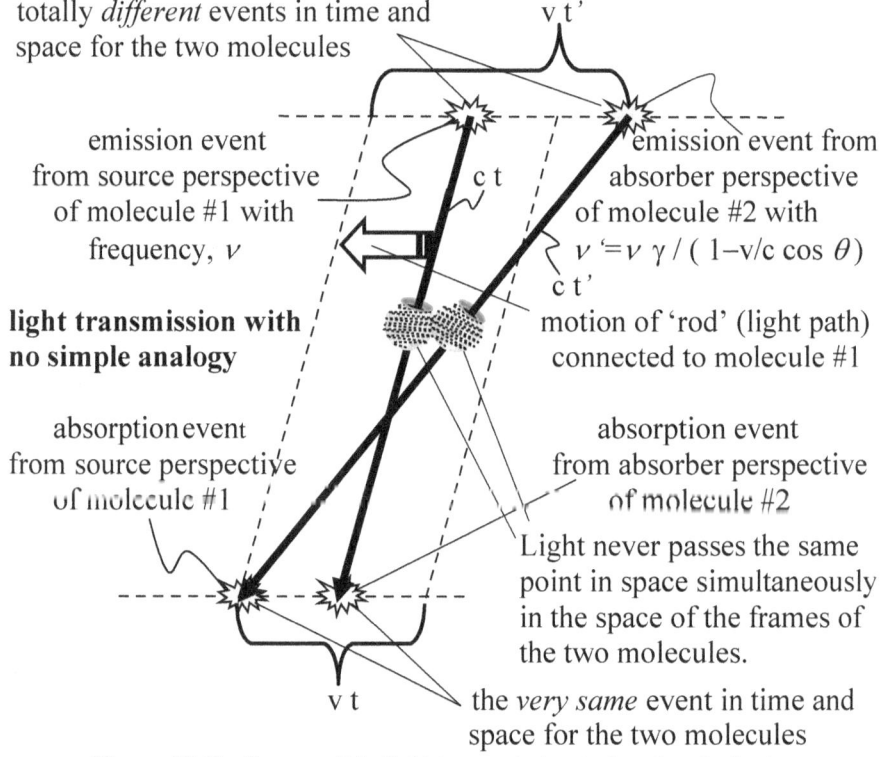

Figure 17.10: Irreversible light transmission in 'modern' physics

To demonstrate the way in which the more modern concepts of light emission should be applied, events can again be perceived from two unique perspectives, that of molecule #1 and that of molecule #2. So in contrast to the single panel a shown in figure 17.6 with the perspective of molecule #2, we now need to show two (what appear to be) very basically incompatible, perspectives. And... significantly, *the reversed situation cannot occur because of the quantum constraint.*

Again in panel a1 of figure 17.11 we illustrate the initial situation as though molecule #1 is at rest. The velocity $\mathbf{v}_{21}$ of molecule #2 is just the relative velocity between molecules #1 and #2. In panel a2 we show the perspective of molecule #2, which we must use as described earlier.

The maximum angle θ with respect to the direction of relative motion from the perspective of molecule #1 is determined from the formula plotted in figure 17.3. Given an initial velocity of ½ c, we can select an angle that accommodates molecular interaction, i. e., any angle less than where the curve crosses the ordinate value of unity (note the circle), the upper limit allowed by the constraint. So for purposes of depiction in figure 17.11.a1 we have selected an initial velocity of ½ c and in accordance with the constraint identified above, we have that:

Relativistic Doppler Effect
panel a1 (forward motion – perspective of molecule #1)

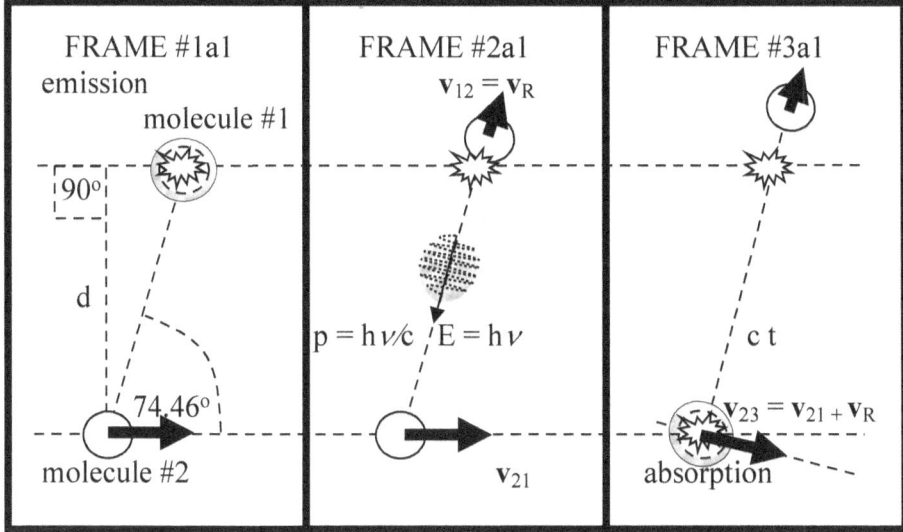

panel a2 (forward motion – perspective of molecule #2)

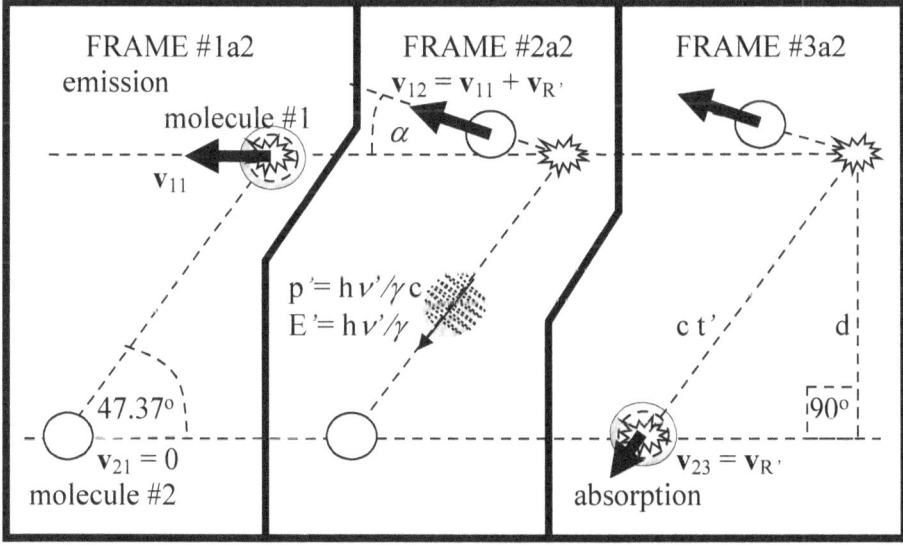

Figure 17.11: Different perspectives of an interaction in relativistic physics

cos $\theta \geq 0.26795$

This limit corresponds to an angle of 74.46° for which the ratio of the change in frequency of the exchanged photon will be:

$v'/v = (1 - v^2/c^2)^{1/2} / (1 - v/c \cos \theta) = (0.866) / (0.866) = 1.000$

Thus, from molecule #1's perspective there is just the amount of energy released to effect the interaction of elevating the energy of molecule #2 to the level vacated in molecule #1 if there were no recoil. However, from the perspective of molecule #2 the frequency is greater than that.

To determine the perspective of molecule #2 we must take into account the relativistic aberration of the angle of approach for the two molecules. Because of the aberration of light (any electromagnetic radiation), the angle from the perspective of molecule #2 differs from the perspective of molecule #1. The relativistic aberration formula appropriate to this case is:

cos $\theta' = (\cos \theta + v/c) / (1 + v/c \cos \theta)$

So the angle $\theta = 74.46°$ from the perspective of molecule #1 is altered to obtain the corresponding angle from the perspective of molecule #2:

cos $\theta' = (0.26795 + 0.5) / (1 + 0.5 * 0.26795) = 0.6772$

Thus, from the perspective of molecule #2 the angle of approach will be decreased to $\theta' = 47.37°$.

Now we will look at the conservation of energy and momentum from both perspectives; in this case we will use the concepts of 'modern' physics and again do that on a frame by frame basis.

Initially, from molecule #1's perspective, it is stationary with respect to molecule #2 that is in motion; all of the kinetic energy and momentum is vested in molecule #2. But again we must adopt the perspective of molecule #2 so the emission derives from a relatively moving source. Molecule #1 emits a photon at 74.46° to the horizontal from its perspective, thereby picking up recoil momentum in the opposite direction. The amount of momentum picked up by the 'stationary' molecule emitting a photon of frequency v was illustrated in discussion of figure 16.4. We have for the equal and opposite momenta:

h $v / c = m_o v_R / (1 - v_R^2/c^2)^{1/2}$

So from this initially relatively-stationary perspective, we obtain for the recoil velocity, v_R:

$$v_R = H / (1 + H^2/c^2)^{1/2}$$

where we again define:

$$H = h \, \nu / m_0 c$$

Again we state that photon recoil velocities are extremely small in comparison to typical velocities of molecules in a thermal gas:

$$H = 1.323 \times 10^{-14} \nu = 3.98 \times 10^{-4} / \lambda$$

Thus even for Lyman-alpha photon emission from a hydrogen atom we would have the small recoil velocity of $v_R \approx 32.63$ cm/sec. Even at room temperatures for which the most probable speed is on the order of 2×10^5 the recoil velocity would be smaller by a factor of 10^{-4}.

But to proceed, in order to take into account the angle of the recoil, and since we are partitioning momentum into horizontal and vertical components, we need to employ the proper trigonometric relations. In addition we must take into account the aberration of the recoil angle to obtain the perspective of molecule #2. Finally, to determine the angle α of the final velocity of molecule #1 from the perspective of molecule #2, consider the geometrical relations shown in figure 17.12:

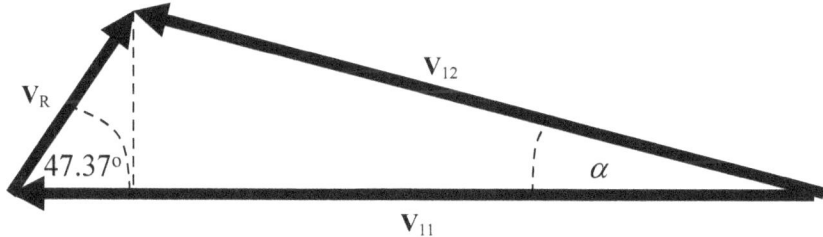

Figure 17.12: Impact of recoil in relativistic treatment

$$v_R \sin 47.37° = v_{12} \sin \alpha$$

$$\sin \alpha = 0.7357 \, v_R / v_{12}$$

$$v_R \cos 47.37° = v_{11} - v_{12} \cos \alpha$$

$$v_R^2 = v_{11}^2 - 2 \, v_{11} \, v_{12} \cos \alpha + v_{12}^2$$

$\cos \alpha = (v_{11}^2 + v_{12}^2 - v_R^2) / 2 \, v_{11} \, v_{12}$

So let us look at the frame-by-frame analysis with velocity v_{21} and frequency v given for the initially 'stationary' frame of molecule #2. Again, the law of the conservation of energy assures us that the total energy E_{TOTAL} of all components before and after the interaction described graphically in figure 17.11, panel a2 must remain the same. Likewise, the law of the conservation of momentum assures us that the momenta p_H and p_V before and after each frame described graphically above also remains unchanged.

FRAME #1a1:

$E_{TOTAL} = m_o \, c^2 + \gamma_{11} (m_o \, c^2 + h \, v)$

horizontal momentum:

$p_H = - \, m_o \, v_{11} \, \gamma_{11}$

vertical momentum:

$p_V = 0$

FRAME #2a1:

$E_{TOTAL} = m_o \, c^2 + m_o \, c^2 \, \gamma_{12} + h \, v \, (1 - v_{11} \cos 47.37^\circ) / \gamma_{11}$

horizontal momentum:

$p_H = - \, m_o \, v_{12} \, \gamma_{12} \cos \alpha - h \, v \, (1 - v_{11} \cos 47.37^\circ) \cos 47.37^\circ / c \, \gamma_{11}$

vertical momentum:

$p_V = m_o \, v_{12} \, \gamma_{12} \sin \alpha - h \, v \, (1 - v_{11} \cos 47.37^\circ) \sin 47.37^\circ / c \, \gamma_{11}$

FRAME #3a1:

$E_{TOTAL} = m_o \, c^2 \, \gamma_{12} + (m_o \, c^2 + h \, v) \, \gamma_{22}$

horizontal momentum:

$p_H = - \, m_o \, v_{22} \, \gamma_{22} \cos 47.37^\circ - m_o \, v_{12} \, \gamma_{12} \cos \alpha$

vertical momentum:

$p_V = - \, m_o \, v_{22} \, \gamma_{22} \sin 47.37^\circ + m_o \, v_{12} \, \gamma_{12} \sin \alpha$

And the original (differently perceived) energy and momenta situation is restored.

Solving the equations yields:

$\gamma_{11} (m_o \, c^2 + h \, v) = m_o \, c^2 \, \gamma_{12} + h \, v \, \gamma_{11} (1 - v_{11} \cos 47.37^\circ)$

$$\gamma_{12} = \gamma_{11} \left(1 + (H/c) \left(1 - 0.6773 \, v_{11} \right) \right)$$

$$= 1.1547 \left(1 + 0.66135 \, H/c \right)$$

$$v_{12} \approx 0.49999 \, c$$

$\sin \alpha = 32.63 \times 0.7357 / 0.49999 \, c \approx 1.6 \times 10^{-9}$ rad. $= 9.17 \times 10^{-8}$ deg.

So, of course the scale is far wrong in figure 17.12. This scale misrepresentation will be true for virtually all viable case for which the mass of the hydrogen atom and frequency of the Lyman-alpha emission line are used. See the table below:

initial relative velocity	recoil parameter H (cm/s)	emission angle constraint	aberrated emission angle	diverted path angle (degrees)	diverted relative velocity
– 0.1 c	32.63	87.13°	81.42°	$\approx 1.1 \times 10^{-7}$	$\approx – 0.1$ c
– 0.5 c	32.63	74.46°	47.37°	$\approx 9 \times 10^{-8}$	$\approx – 0.5$ c
– 0.9 c	32.63	51.19°	12.55°	$\approx 2.6 \times 10^{-10}$	$\approx – 0.9$ c

Furthermore, this situation is even more extreme for more massive atoms and molecules and smaller (more reasonable) frequencies of the radiation. But significantly in this case and for smaller and larger relative velocities, the conservation laws can be satisfied. Thus modern physical concepts do not deny the viability of submicroscopic interactions. There are indeed a full spectrum of viable cases of approaching relative velocities that abide the constraints of figures 17.3 and 17.4. However, even more significantly, none of these interactions is reversible.

the nature of submicroscopic irreversibility

In virtually every submicroscopic interaction that is mediated by a photon there is a transfer of energy and momentum between molecules that reduces their relative velocity. Ultimately they are receding rather than approaching each other and there are no more opportunities of interaction. The impact of this is similar to the irreversible inevitability of entropy that drains energy from the high energy sources of a thermodynamic system that had increased the general wealth of energy of the system. In the reversible classical analyses there are no irreversible changes in the relative status of two interacting molecules. However, in the case of analyses with 'modern' physics, there are.

The possibility of what might occur in panel a2 of figure 17.11 from the perspective of molecule #1 is similar to what was envisioned to have transpired in panel a of figure 17.6. But what is seen to transpire from the perspective of molecule #2 is quite different. Since molecule #2 will ultimately witness (absorb) the photon emitted from molecule #1 from an aberrated angle that (unlike in classical theory) is very basically incompatible with any place or time of emission and/or absorption events according to clocks (if there be such) of molecule #1. The wavelength will be shorter from his perspective, producing a higher frequency and therefore recoil velocity than was the case for the classical analysis.

Let us address prospective ramifications of reversing time or the velocities from both perspectives. From the perspective of molecule #2 the angle that molecule #1 would see the radiation coming from molecule #1 if their relative velocities had stayed the same can be determined from the reverse aberration formula as follows:

$$\cos \theta \ = (\cos \theta' - v/c) / (1 - v/c \cos \theta') = 0.26795$$

Now we obtain $\theta' = 74.46^\circ$ and there is basic agreement on what angles will be realized. Of course their relative velocity is not the same after the photon has been absorbed, but nearly so. And in any case once the velocities have been reversed in this exercise, there will be recession of the two molecules rather than approach so the quantum effect totally precludes any reverse interaction.

But although the aberration and Doppler formulas are symmetrical with velocity reversal and the conservation laws would be satisfied, the net result is that there can be no reversibility of the exchange. By the reversal of velocities we would have the two molecules receding from each other rather than approaching. Receding molecules cannot (without coincidental shifting to another different discrete energy level) exchange quantized photons, because they are precluded by quantum restrictions on the amount of energy an atom or molecule can accept in an energy exchange. Of course this does not mean that the original interaction is precluded or that there are not similar interactions that can proceed in the opposite direction. But – counter to virtually every commentary to the contrary – reversibility does *not* occur on all (or in fact virtually *any*) mediated interactions that take place at the submicroscopic level of reality. Irreversibility originates here!

There is indeed some similarity to the elastic collisions described by Boltzmann. But whereas all of the elastic collisions are examples of reversible behavior, virtually none of the mediated exchanges are reversible. And in fact as the repeated exchanges reduce the relative velocity to zero, the viability of further exchanges between the two molecules ceases.

necessary extensions to photon mediation analyses

In the preceding analyses and the previous chapter we dealt with a homogeneous medium with all particles having essentially the same rest mass m_o or M_o depending on whether they were in their high or low energy state associated with radiation of frequency v. When we applied numbers to the observable behavior, we used the mass of hydrogen and the frequency of the Lyman-alpha emission line as an example. This is a worst case as far as the magnitude of recoil velocities would be concerned. So admittedly we have not taken into account the broad diversity of the interactions that actually occur in a thermodynamic system at the submicroscopic level. But the conclusions with regard to behavior at this level are unchanged by this didactic simplification.

Viable photon interactions involve like particles and like discrete energy exchanges in virtually all cases. Exceptions involve fairly extreme Doppler shifts of a photon into the domain of a different specific spectral line or perhaps even a different atom type or even into the freeing of electrons by photons of high enough energy to be in a continuum of radiation above spectral lines. For example if energy associated with the radiation frequency is greater than the depth of the potential energy well associated with the bound electron, then the electron and nuclei will disassociate with the photon momentum and energy all being taken up in that case by the freed sub-particles. These situations are primarily only encountered in hot plasma gases but still within the realm of thermodynamics as we will see.

Perhaps the more significant extension that needs to be considered is scattering of radiation by multiple particles without absorption taking place. This is yet another type of submicroscopic interaction not taken into account by either Boltzmann's or specifically by Einstein's analyses.

Part IV

18: Electromagnetic Scattering

"It goes without saying that thermalization of electromagnetic radiation requires scattering of the radiation by matter."[19]

We have demonstrated that it is by transmission of electromagnetic radiation between molecules of a gas that irreversible behavior is first encountered at the submicroscopic level. But we need to understand the nature of all the interactions of radiation with the charges present in any non-vacuous medium. Up to this point we have dealt exclusively with intermolecular energy exchanges with nothing to interfere between the two molecules. Of course what Einstein denominated 'stimulated' emission is an exception of sorts that is addressed more specifically with regard to electron resonance with passing radiation.

We have come to understand the qualifications involved in direct photon mediation between two molecules, as in radiation being emitted by one molecule, traveling directly to another isolated molecule without interference where it is absorbed. We have discussed criteria for these in the previous chapters including the requirement for precise balance of emitted and absorbed energy and the associated fact that only those molecules that are approaching each other can interact in this way. So it requires a unique set of molecules to be eligible for such interactions.

But even with a homogenous gas with all molecules of the same type, there is a vast range of frequencies of radiation that could be exchanged, with each side of the interchange possessing and being receptive to the same frequency. So it would be unusual if two closest neighbor molecules were compatible for such an exchange. Emitted

---

[19] Bonn (2011) p. 571.

radiation will most often pass by the vicinity of other molecules that are not eligible to absorb it but have electronic charges that will be affected by it. A complete vacuum throughout substantial regions of space is never realized. So we must consider the effect of extraneous molecules and specifically the associated electrons on transmission of radiation.

To the extent that we have neutral molecules, i. e., temperatures are low enough that there is no considerable disassociation of electrons and nuclei, electrons will be bound to nuclei of the atoms within molecules. But these electrons nonetheless respond to electromagnetic fields in the vicinity of the molecules causing minor displacement of the centers of the positively charged nuclei and the negatively charged electron clouds surrounding them. A 'dipole moment' is thereby established, producing an electromagnetic field of its own. See figure 18.1.

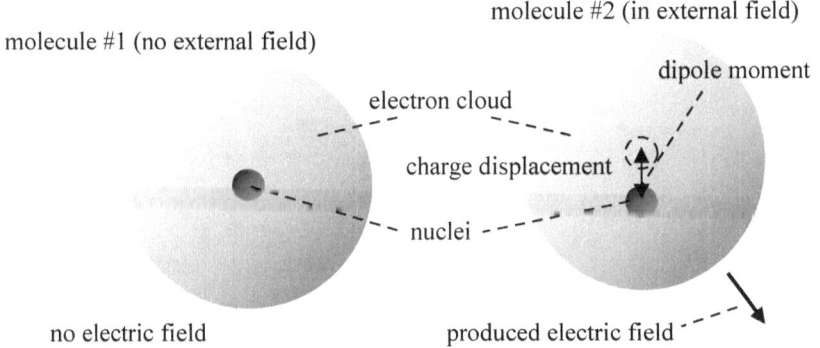

Figure 18.1 The effect of external electric field on molecular charge distribution

In situations when there are electric charges (even those in neutral molecules as shown in the figure) between the emission and absorption of radiation by two primary interacting molecules, there are additional considerations. The fields produced by dipole moments along the propagation path affect the transmitted photon. Electromagnetic field expressions and the physics of scattering are embodied in the Lorentz-Lorenz formulas that determine the extent of these effects.

It should be noted that with regard to scattering phenomena we will most often treat electromagnetic theory from a perspective of classical physics. As we mentioned earlier, quantum theory involves two phases of light transmission, a first phase involves a wave function and the second phase involves collapse of that wave function upon absorption of a photon. In scattering with which we will be concerned absorption is not involved, so in this case the wave aspect of the wave/particle duality of light is what we will address.

A significant aspect of the scattering of light is that by which a dispersive medium effectively replaces photons as a part of a coherent forward scattering process. This process supports the imaging of objects and, in fact, everything we observe through our atmosphere has resulted from photons having been replaced at millimeter intervals. The topic is not generally discussed in much detail because it has long been presumed that the wavelength of transmitted radiation is not affected by this process, etc.. So the effects through which photons are continually replaced have not seemed all that important other than in observations made specifically to refine constraints on the constancy of the velocity of light propagated through a vacuum.

Nonetheless, we will describe this forward scattering process in some detail in the next chapter because, however negligible the effect on a given transmission, we will demonstrate that the effect accumulates and is irreversible. Analyses require determination of the value of what is referred to as the 'extinction interval' after which the originally emitted photon will have been replaced by a similar one but emanating from scattering electrons. This distance after which incident photons are replaced by 'cloned' versions of themselves is inversely proportional to the wavelength of the radiation. There is a minute lengthening of wavelength at each replacement of a photon by scattering events that result in the transfer of momentum (and energy) to the intermediate material substance. The amount of lengthening of the wavelength of the photon is inversely proportional to the original wavelength of the incident photon.

simplified discussion of electrodynamic wave equations

We will not go into any very exhaustive explanation of Maxwell's equations of electrodynamics. Readers are expected to appreciate, however, that all radiation is propagated in accordance with those laws, only slightly modified in most cases by quantum considerations. Furthermore, scattering phenomenon that is specifically the subject of interest here can be explained in terms of electromagnetic fields defined for use in Maxwell's equations. We will use monochromatic incident waves (a single wavelength λ) in describing scattering phenomena. This simplifying assumption is actually valid for broader generalization since all radiation can be represented as linear summations of just such monochromatic waves. The solutions to Maxwell's equations for the incident wave are of the form:

$$E_i = E_o \, e^{\, i \, (\, \omega \, t - 2 \, \pi \, \mathbf{n} \, z / \lambda \,)}$$

where the electric field E_i is in general a vector, but we will define it here as a scalar along the x direction in Cartesian coordinates with propagation along the z direction. E_o is a constant, i is the square root of minus one, ω is the radial frequency of the radiation, **n** is the index of refraction, and λ the wavelength of the radiation. We have

$$\omega = 2\pi \, \nu = 2\pi \, c \, / \, \lambda$$

We will use plane waves as a basis for our discussions, which acknowledges that as electromagnetic radiation propagates to greater and greater distances from a relatively localized source, the wave front can be approximated by a *plane wave* with ever increasing accuracy because limited surfaces on very large spheres are very nearly planar. This simplifying assumption is justified to extremely high precision for radiation that propagates to a distance that is considerably greater than the average *extinction interval.*[*]

forces imposed on individual material charges

The effect of electromagnetic radiation on material substances is to force the acceleration of individual constituent charges of which the substance is comprised. The same basic phenomena occur whether the constituent charges are more or less 'bound' within atoms, molecules, crystalline structures, or are 'free' as in diffuse ionic plasma. The accelerations of these charges are what cause the emanations of 'secondary' radiation of basically the same frequency originating at the individual constituent charges. This phenomenon is associated with what is called 'scattering'. It is illustrated for an individual charge in figure 18.2 below. In panel a, a charged electron is forced to oscillate at the frequency of the incident radiation with dynamics as shown in panel b. The electric field emanating from the charge will exhibit, in addition to an ineluctable radial field that is diminished quickly as the inverse square of distance, an oscillating transverse component, labeled E_{Sx} in panel a, that diminishes less rapidly and will not be cancelled like the outward component will. The transverse component will be oppositely directed from that of the incident field, tending to cancel it as shown in panel c.

We will derive many of the primary dispersion formulas applicable to thermodynamic systems. These formulas derive from just the kinds

---

[*] This is the distance after which the properties of the medium determine propagation characteristics. The concept of 'extinction' and particularly the term 'extinction distance' differ with some treatments that use the term for absorption distance..

of considerations shown in figure 18.2. Then we will apply the dispersion formulas to explore the effects of scattering on thermalization whereby kinetic and radiational energy distributions become synchronized. Finally, the applicability of these formulas will be re-examined to determine the impact of high temperature electrons for which relativistic effects are encountered in intergalactic plasma.

panel a: impact of incident radiation on a charged particle

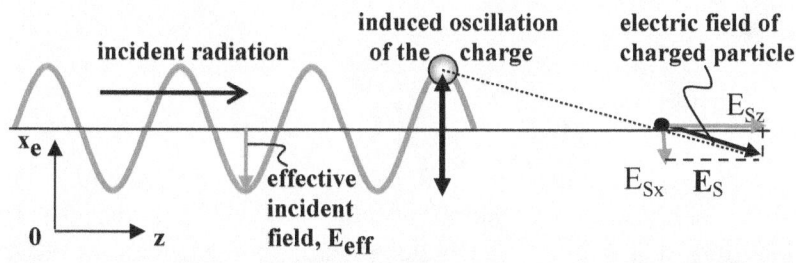

panel b: charged particle dynamics of a single cycle

acceleration $= \text{force} / m_e = (e/m_e)E_{eff}$
$\qquad\qquad = dv_e/ dt = d^2 x_e/ dt^2$
velocity $\qquad = v_e = m_e\, d\, x_e / dt$
position $\qquad = x_e$

time $\rightarrow$

panel c: superposition of fields at a distance from the charge

incident radiation field, E_i
field due to charge displacement, E_{Sx}
net effective field, E_{eff} at a distance
from the displaced charge

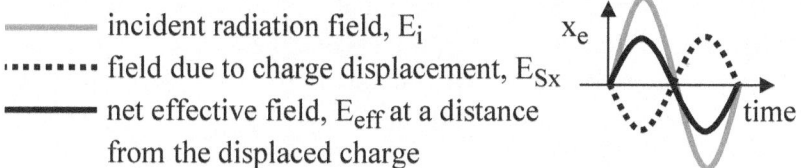

Figure 18.2 Impact of incident radiation on a charged particle

Incident electromagnetic radiation E_i together with induced fields of other charge displacements will cause individual electronic charges throughout the medium to experience a Lorentz force that will produce the displacements of the charge. This force is given by:

$$\mathbf{F}_e = e\, \mathbf{E}_{eff}$$

where e = 4.8×10^{-10} statcoulombs (in cgs units that we will use here) is the individual electronic charge, v_e is the electron's velocity with respect to an observer of the scattered radiation. We will treat only the contribution of the electric field here both with respect to induced polarization and in total; the effects of the magnetic field could be treated analogously, but they can be shown to have no unique significance for most substances in regions where there is no appreciable static magnetic field.

The effective electric field is related to the incident electric field E_i as follows:

$$E_{eff} = E_i + (4\pi/3) P$$

The presence of the polarization, **P**, must be taken into account because the force realized on a charge at any point in a substance is a combination of that due to an incident radiation field and an *induced* polarization field that opposes it. This induction process is illustrated in figure 18.2.

The induced field is produced in surrounding regions of a substance because of an induced polarization of charges in the within that substance – positive charges will be 'induced' to migrate in one direction, negative ones in the other. **P** is known as the polarization of the medium.

We also consider only displacement of the negatively charged electrons and not that of positive ions such as the nuclei of molecules. The much greater mass of the positive nuclear ions (even when they are primarily single protons) will preclude their accelerations becoming appreciable relative to those of the much less massive electrons at frequencies with which we will be primarily concerned.

Each electron will experience, in addition to the Lorentz force, restoring and damping forces representing a medium's ability to restore the locally altered equilibrium and its tendency to dissipate energy. The resulting displacements will be along the x-axis under the assumptions of isotropy and homogeneity of the medium for incident monochromatic radiation proceeding along the z Cartesian coordinate direction. The formulation of these forces in the usual F = m a (mass time acceleration) on a single electron is typically represented by the following differential equation of classical electrodynamics:

$$m_e \, d^2x_e/dt^2 = e \, E_{eff} - k \, x - g \, dx_e/dt,$$

where the second and third terms on the right correspond to a proportional restoring force and a velocity-dependent damping force associated, respectively, with the harmonic restoring force on charges about a stable neutral position and an absorptive effect typically associated with collision probabilities while the incident wave is interacting with the electron, effecting joule heating. These coefficients will determine the magnitude of an inherent resonant frequency of the scattering electrons and the absorption properties to be expected of the medium as a whole. Solution is considerably simplified by addressing only a single resonant frequency; this simplification is appropriate for a fully homogeneous medium in Maxwell-Boltzmann equilibrium.

Solving the previous differential equation results in a displacement formula:

$$x_e = (e/m_e) \, E_{eff} / (\omega_o^2 - \omega^2 + i \, \gamma \, \omega), \text{ where}$$

$$E_{eff} = E_o \, e^{\,(i \, \omega \, t - \frac{1}{2} \gamma \, t)}$$

x_e is a displacement that is in direct response to the incident radiation. The resonant radial frequency of the medium is ω_o.

Figure 18.1 illustrates the mechanism whereby the *effective* field is augmented by these displacements caused by the incident electric field. At some distance removed from the electron that is affected by the incident radiation field, the electronic charge produces an electric field directed away from the charge. There will be a component of this field E_s that is along the direction of the incident radiation and a component at right angles to it as shown. The component that is along the direction of propagation will diminish with distance from the electron according to the usual coulomb inverse square law. However, the transverse component of the field will diminish only as the inverse first power of distance. Thus, after any very appreciable distance E_s will itself comprise transverse radiation that is one quarter wavelength behind that of the incident electromagnetic wave and hence will reduce the magnitude of the incident wave accordingly.

These scattering fields from individual electrons in the medium will accumulate by coherent constructive reinforcement until eventually their combined effect is to totally nullify and replace the incident radiation altogether.

Both ω_o and γ play key roles in determining the effects the medium will have on the incident radiation. From the above equations, we obtain the following:

$\omega_0{}^2 = k / m_e - g^2 / 4\ m_e{}^2$, and $\gamma = g / m_e$, such that,

$$\gamma^2 / 4 - (k / m_e)\ \gamma + \omega_0{}^2 = 0$$

This relation between k, γ, and ω_0 is shown in figure 18.3. It expresses relationships between the force constants and derived parameters that determine scattering behavior. These important parameters value are determined by specifics of the scattering process.

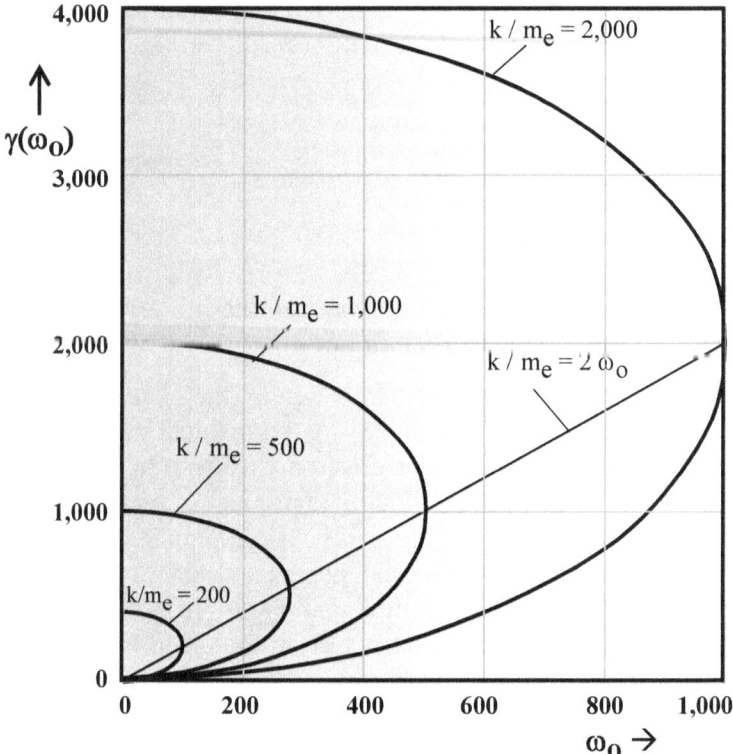

Figure 18.3: Relationship between resonant frequency and absorption constants

dispersion formulas

The induced *dipole moment* illustrated in figure 18.1 is a reaction to the incident electric field; it is also oriented along the x axis in diagrams of figure 18.2 appropriate to an isotropic medium with which we primarily concern ourselves. The reaction force on a single electron resulting from this displacement is given by:

$$\mathbf{P_e} = e\ \mathbf{x_e} = (e^2/m_e)\ \mathbf{E_{eff}} \ / \ (\omega_0{}^2 - \omega^2 + i\ \gamma\ \omega\ )$$

where P_e is the contribution of a single electron to overall polarization, **P**, of the entire medium at a given point in space.

The effect on the incident radiation of the interaction with the electronic charges is that overall transmission characteristics may be significantly altered by refraction, diffraction, and/or absorption. This dispersion is also associated with a wavelength dependent propagation velocity of the on-going radiation.

In the simplified Lorentz force equation presented above, the introduction of scattered electric fields resulting from the induced polarization effects on the incident electric field realized throughout the medium is associated with a property called the *refractive index* or *index of refraction*, **n** of the medium. The quantity **n** is in general a complex quantity (i. e., possessing both a 'real' and an 'imaginary' component) that determines propagation velocities in the medium as follows:

$$\mathbf{n} = c / \mathbf{b},$$

where **b** is the *complex* speed of light in the medium and c is again the universal speed of light in a vacuum. The *phase velocity* of radiation in the medium is represented by the real part of **b** and the *group velocity* by its *norm*. We will continue the dispersion formula derivations using **n** rather than **b** as is usual in such discussions. The imaginary part of **n** determines the absorption properties of the medium as we will show.

In the absence of polarization effects in the medium, we would have had from electromagnetic theory that the induced electric field (the *electric induction*) in the medium, $\mathbf{D_r}$, would be determined as:

$$\mathbf{D_r} = \mathbf{E_i}$$

However, in a polarized or polarizable substance, this becomes:

$$\mathbf{D_r} = \in \mathbf{E_i}$$

where $\in$ is the *dielectric constant* of the medium, defined as:

$$\in \equiv \mathbf{n}^2$$

For an isotropic medium $\in$ is a scalar so that it can be assumed that the vector quantities $\mathbf{E_{eff}}$, **P** and $\mathbf{E_i}$, employed earlier, as well as $\mathbf{D_r}$ are all aligned along the same direction. This will always be the case in s

233

homogeneous isotropic medium, so scalar ratios are meaningful. We will assume these simplifying, but usual, assumptions. In this case:

$$\mathbf{D_r} = \mathbf{E_i} + 4\pi \, \mathbf{P},$$

and since vectors are aligned, we have a scalar result:

$$\epsilon = 1 + 4\pi \, P/E_i$$

Therefore, we have that:

$$\mathbf{E_{eff}} = (\, n^2 + 2 \,) \, \mathbf{E_i} \, / \, 3$$

The presence of the factor of 1/3 in the effective electric field equation is known as the *Lorentz-Lorenz correction*; the justification for this factor is somewhat involved so it will just be assumed in the present discussion. See for example Jackson (1962), Ditchburn (1963) or any other detailed text on electrodynamics. And the polarization field will be given by:

$$\mathbf{P} = (\, n^2 - 1 \,) \, \mathbf{E_i} \, / \, 4\pi$$

The polarization field, **P**, required in the determination of the effective electric field realized at points within the medium is defined as the dipole moment per unit volume as follows:

$$\mathbf{P} \equiv \rho_e \, \mathbf{P_e}$$

where ρ_e is scattering electron density. By substitution for **P_e** from the equation derived above, we obtain the formula:

$$\mathbf{P} = \rho_e \, (e^2/m_e) \, \mathbf{E_{eff}} \, / \, (\omega_0^2 - \omega^2 + i\,\gamma\,\omega\,)$$

the Lorentz-Lorenz formula

By further substitution of the derived values we obtained above for P and E_{eff} into the preceding formula, we obtain a formula for the index of refraction as follows:

$$(\, n^2 - 1 \,) \, / \, (\, n^2 + 2 \,) = (4\pi/3) \, \rho_e \, (e^2/m_e) \, / \, (\omega_0^2 - \omega^2 + i\,\gamma\,\omega\,)$$

This is the *Lorentz-Lorenz formula* for the index of refraction applicable to media generally exhibiting a single predominant resonant frequency. For cases where the index of refraction is close to unity as it will be for thermodynamic gases that we will consider, the left-hand side of the previous equation becomes approximately 2 (**n** − 1) / 3. This in turn results in the usual determination that is valid to a very high degree of accuracy, for which the *dielectric susceptibility*, (**n** −1), is assessed as follows:

n − 1 ≈ 2π (ρ_e e^2/m$_e$) / (ω_0^2 − ω^2 + $i\,\gamma\,\omega$)

The index of refraction, being complex, can be expressed in the form:

n = *Re*(**n**) + *i* *Im*(**n**)

where *Re*(**x**) and *Im*(**x**) refer respectively to exclusively real and imaginary components of a complex argument, **x**. From the formula for the dielectric susceptibility above, we obtain:

Re(**n**) = 1 + 2π (ρ_e e^2/m$_e$) (ω_0^2 − ω^2) / ((ω_0^2 − ω^2)^2 + ($\gamma\,\omega$)^2)

Im(**n**) = − 2π (ρ_e e^2/m$_e$) $\gamma\,\omega$ / ((ω_0^2 − ω^2)^2 + ($\gamma\,\omega$)^2)

These functions are plotted in figure 18.3 for nominal parameter values of ω_0= γ = 87. The forms of these two curves are quite general, but not without quite extreme domain peculiarities. See for example the discussion by Bonn.[20]

In exploring the ramifications of these formulas, it is clear that behavior will be markedly different if the frequency ω of the radiation that is propagated through the medium is higher or lower than the resonant frequency, ω_0 that is a characteristic of the medium itself. In all cases we will discuss in this volume it will definitely be higher.

The topic of the next chapter will be forward scattering in a homogeneous isotropic medium. It involves many constituent electrons where 'coherent' positive and negative interference of electromagnetic fields plays an all-important role. Individual contributions to the polarization field symbolized earlier in figure 18.1 accumulate because of constructive interference that only applies to the extent that the

---

[20] R. Bonn, Cosmological Effect of Scattering in the Intergalactic Medium, Vaughan Publishing, Seattle, pp. 89-94, (2011)

electromagnetic fields that interfere are indeed 'coherent'. This is a concept that we will discuss in more detail. The superposition principle that is the basis of the accumulation process guarantees that scattering effects can be handled independently of ongoing incident radiation. We will look into mechanisms whereby, rather than just gradually absorbing incident radiation, intermediate scattering electrons actually effect replacement of the incident radiation by similar radiation having a different phase and with a slightly reduced speed because of the altered path. The process involves the summation of the individual effects from the various scattered electromagnetic fields originating at the individual intermediate electrons to ultimately effect replacement of the incident radiation.

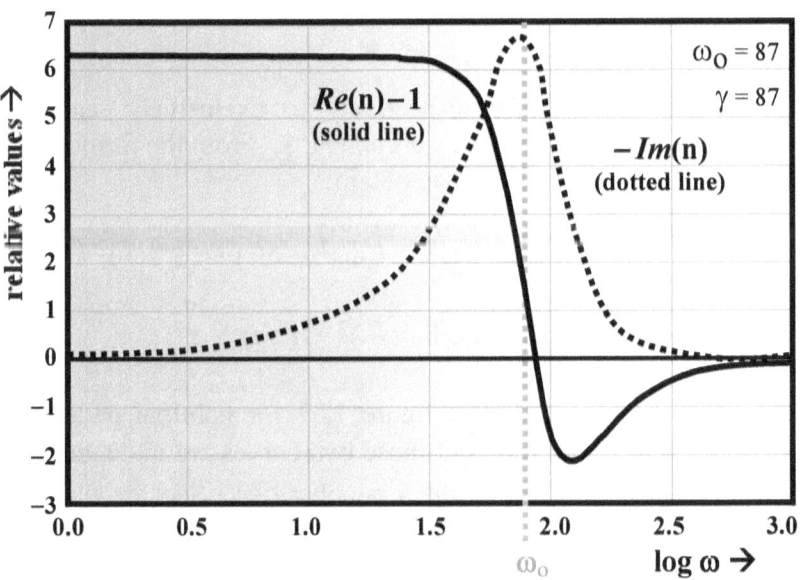

Figure 18.3: Real and imaginary components of complex index of refraction

characterization of the scattered electromagnetic field

Let's consider the field emanations of induced electron oscillations from the individual constituents of the scattering medium. The radial component of these resulting field emissions diminish as the inverse square and higher order powers of the distance from the constituent electrons as shown earlier. The radial component will, therefore, be negligible after a modest distance with respect to the wavelength of the incident radiation. This is in contradistinction to the transverse component, one of whose terms diminishes as only the inverse *first power* of distance. This component of scattered electric fields radiated as transverse radiation by an accelerated electron will be oriented in the opposite direction (i. e., *out of phase*) from that of the incident radiation field that induced it. This is shown in panel c of figure 18.2. Refer to pages 523 through 529 of Ditchburn (1963) or similar text for an in-depth discussion of the radiating dipole applicable to this analysis, where it is shown that the transverse component of the radiation from a dipole of moment P_e is:

$$E_e(\omega,r) = -(\omega^2/c^2) |P_e| \cos \alpha / r$$

$$= -(\omega^2 / 2\pi \rho_e c^2) | n - 1 | \cos \alpha / r$$

It is assumed that the distance r from the electronic charge is much greater than the wavelength, i. e., $r \gg \lambda$, where r is the distance and $\lambda = 2\pi c / \omega$ is the wavelength of the incident radiation. The angle α is the scattering angle of the radiation with respect to the direction of the propagation of the incident radiation. This *scattering angle* will always be extremely close to zero for coherent forward scattering that we will look into in particular. That is because of the constructive interference in this direction as we will show. These considerations ultimately dictate that there will be cancellation of all scattered fields arriving from outside an angular domain that will be extremely narrow so that we can assume that $\cos \alpha = 1$ and of no consequence to the realized field strength or intensity at distant locations.

With substitution of parameters for the general expression of the index of refraction, the intensity I_e of the radiation will be:

$$I_e(\omega,r) = E_e(\omega,r)^* E_e(\omega,r) / 2$$

$$= E^2_o \{(e^2 \omega^2/m_e c^2)^2 / ([\omega_o^2 - \omega^2]^2 + [\gamma \omega]^2)\} [\cos \alpha / r]^2$$

However, although the field strength and intensity of the individual scattered radiation are relatively insensitive to slight angular and even to small linear distance changes in the location of assessment, this is not the case with regard to the phase of the radiation. Very small differences in propagation distance and angle produce major changes in the phase that are essential to the analyses pertaining to forward scattering that we will address next.

19: Forward Scattering of Incident Radiation

"Of course forward scattering has typically involved nothing other than mundane optical physics to little spectacular effect. However, the intergalactic medium is hardly typical of media that have been studied in the laboratory."[21]

We have mentioned briefly that the wave function aspect of the wave/particle duality of radiation in quantum physics applies to the 'cloning' of photons that are propagated through a non-vacuous ensemble of molecules. The photon (particle aspect) only really applies once absorption of the photon occurs. A wave function persists throughout space until it ultimately collapses when absorption occurs. With scattering we are dealing with that intermediate realm that was so hard to characterize in our attempts at phases of conservation equations. We are dealing with continuous waves that interfere with each other.

coherent radiation from separate scattering events

If radiation from separated sources is to interfere, the wave functions from each source must be similarly polarized and pass the same observation point within the *coherence length* of individual wave functions as shown in figure 19.1. In the figure, r'_i and r'_j represent the distances of two scattering electrons e_i and e_j from the location of an arbitrarily situated coordinate system. The coherence length is the effective linear dimension of a photon; it is on the order of 10^7 times the radiation wavelength. For arrival separations larger than this amount, an entire photon of radiation from one source will have passed

---

[21] Bonn ((2009) p. 577

a given location before the other photon arrives. In that case neither positive nor negative interference phenomena will occur. The intensity of radiation at the location will just be the simple result of the accumulation of separate photons from the two sources over a designated period of time. However in thermodynamic gases that we are considering, the density of molecules is such that there is plenty of opportunity for interference.

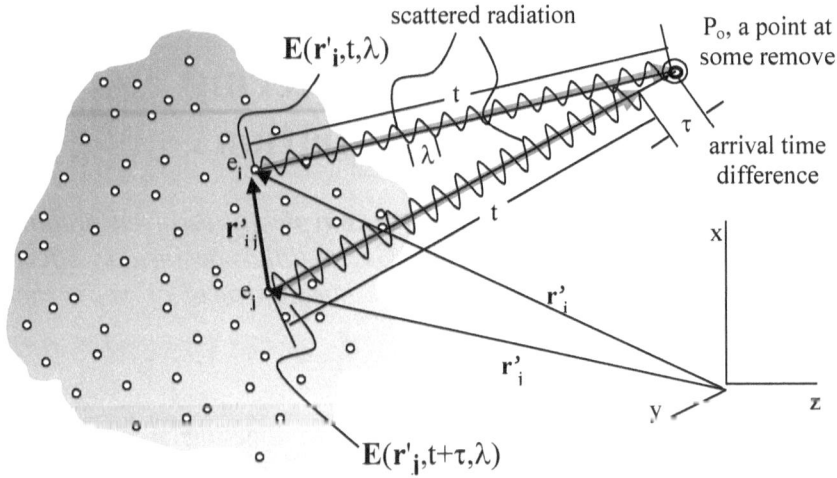

Figure 19.1: Illustrating conditions for distributed source coherence

Christian Huygens proposed in 1670 that the way light propagates is by means of emanations of secondary radiation. This notion is depicted in figure 19.2. Of course he did not have sufficient information about the nature of matter to elaborate the details of these centers of secondary radiation, but precursor ideas of how forward scattering works certainly should be attributed to him. But let us fill in some of those details that could only come much later.

In figure 19.3 we do address the situation of scattering electrons more specifically. In particular, the initiation of their induced radiation is synchronized by having been triggered by the same incident plane wave. The two distances r_i and r_j represent the respective distances of the two scattering electrons e_i and e_j from the location of interest, P_o. Here we illustrate the situation for which the scattered wave functions from two such electrons *do* interfere. In this case, the scattering phenomena are precipitated by oscillations caused by the simultaneous arrival of the incident radiation. Let the angle between the two electrons from the perspective of P_o be α, with t and t+τ the

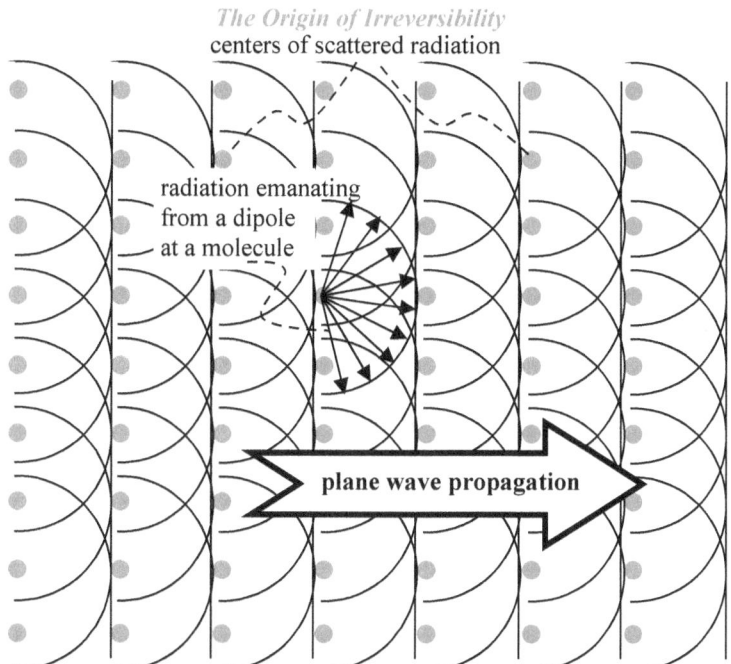

Figure 19.2: Huygens's principle as a precursor of 'forward scattering' theory

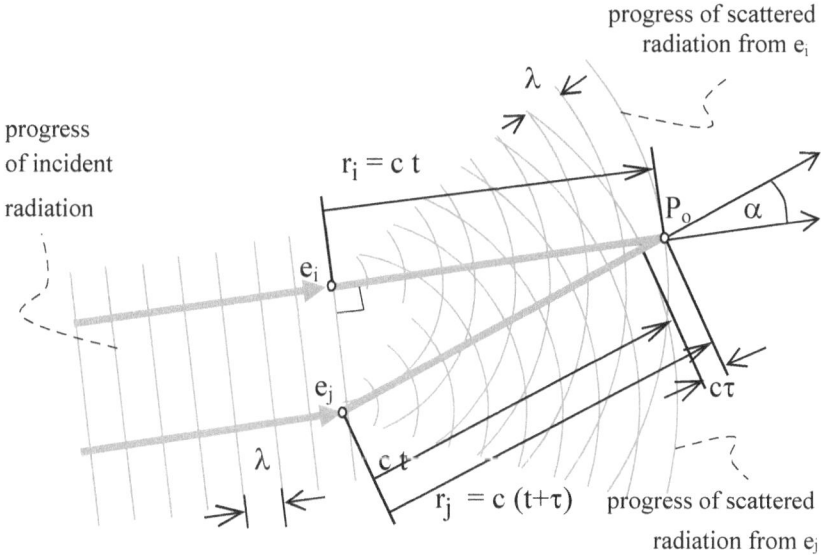

Figure 19.3: Illustrating conditions for coherent forward scattering

respective times of arrival at P_0 for the scattered radiation from the two electrons. The following implications result from the geometrical relations:

$r_j = c (t+\tau) = r_i \tan \alpha$, where $r_i = c\,t$, and:

$\tau = t (1 / \cos \alpha - 1) = t (1 - \cos \alpha) / \cos \alpha$

$= t \sin^2 \alpha / (1 + \cos \alpha) \cos \alpha \approx \frac{1}{2} t \sin^2 \alpha$, for $\alpha \cong 0.0$

With wave phenomena the phase of the waves originating at separate sources affects the net field strength realized at a given location P_o. So, whenever the value of $c\,\tau$ is an integral multiple of λ, the two waves will constructively (additively) interfere. This constraint affects the required separation between two electrons if they are to constructively interfere. These constraints ultimately determine the collaborative effect of intermediate scattering electrons on the radiation observed at a given location.

Consider, for example, the scattered electric fields from the two electrons situated such that they both begin to experience the effects of the incident plane wave radiation at the very same instant as shown in figure 19.3. If $\alpha \cong 0.0$, as we are positing initially, and will confirm as a necessary consequence, then at P_o a distance $c\,t$ from the electron e_i we will have for the total field strength of the scattered radiation at P_o:

$$E_{P_0}(\omega, ct) = E_{e_i}(\omega, ct) + E_{e_j}(\omega, c(t+\tau))$$

$$\cong K(\omega)\, e^{\,i\omega\,(t_o - t)} (1 + e^{\,-i\omega\,\tau}) / c\,t$$

where $c\,t$ is the perpendicular distance to the plane of constant phase of the incident radiation from the electron e_i to the location P_o and $K(\omega)$ includes all those factors in $E_e(\omega, r)$ above that exhibit approximately equal values as long as $\alpha \cong 0.0$:

$$K(\omega) = E_e (e^2 \omega^2 / m_e\, c^2) (\omega_0^2 - \omega^2) / [(\omega_0^2 - \omega^2)^2 + (\gamma\,\omega)^2]$$

Even at large distances with the constraint of a phase difference, $\tau \ll t$, appreciable effects result. The instantaneous illumination intensity of this net scattering radiation at P_o becomes a sinusoidal function of whatever that location-dependent phase difference happens to be:

$$I_{P_0}(t,\tau,\lambda) = [E_{e_i}(ct,\lambda) + E_{e_j}(c[t+\tau],\lambda)] * [E_{e_i}(ct,\lambda) + E_{e_j}(c[t+\tau],\lambda)] / 2$$

$$= (K(\omega) / c\,t)^2 [2 + e^{\,-i2\pi\,c\,\tau / \lambda} + e^{\,+i2\pi\,c\,\tau / \lambda}] / 2$$

$$= (K(\omega) / c\, t\,)^2\, (1 + \cos 2\, \pi\, c\, \tau / \lambda\,)$$

In terms of angular separation of the two electrons from the perspective at P_0 this becomes the following as shown in figure 19.4.

$$I_{P_0}(t,\alpha,\lambda) \cong (K(\omega) / c\, t\,)^2\, [\, 1 + \cos (\, \pi\, c\, t\, [\sin^2 \alpha\,] / \lambda\,)\,]$$

In these equations, of course, the square of the complex factor $K(\omega)$ is:

$$K(\omega)^2 = \{(e^2\, \omega^2 / m_e\, c^2)^2 / (\, [\, \omega_0{}^2 - \omega^2\,]^2 + [\, \gamma\, \omega]^2)\, \}$$

where the general form of the index of refraction is used.

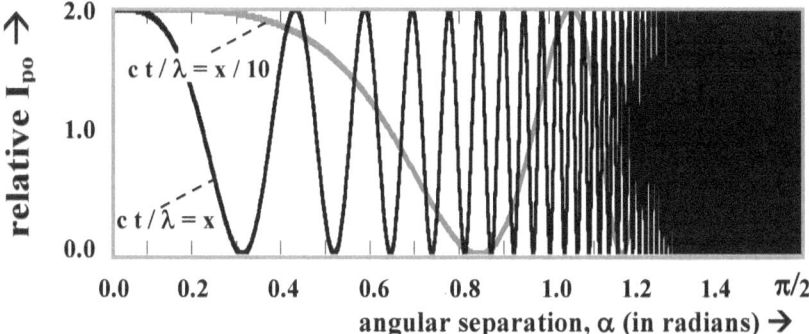

Figure 19.4: Illumination intensity of two separated scattering events

The final term of the trailing factor of the resulting illumination, I_{P_0} is affected by the relative wave phase, τ of the two induced wave functions. Depending upon the angular separation of the two electrons, the resultant effective intensity of the scattering field at P_0 will vary between the extremes of zero and twice the intensity from a single scattering electron as shown. As the distance increases, in addition to the inverse square impact, the predominant central illumination will be narrowed. At appreciable distances it will be very narrow indeed.

Furthermore, rather than just two individual secondary sources interfering, we must consider electrons throughout the entire areas on the planes of incident radiation. Clearly scattered radiation from those electrons closest to the point P_0 will constructively interfere whereas radiation from electrons that are further away will increasingly interfere destructively as well as constructively and therefore not contribute significantly to overall intensity at P_0. In figure 19.5, an annulus of the medium is identified in the thin vertical planar sheet whose thickness is assumed to be much less than the wavelength of the radiation, λ. Throughout the indefinitely-extended sheet incident radiation is assumed to arrive simultaneously from the left instigating scattering by imbedded electrons.

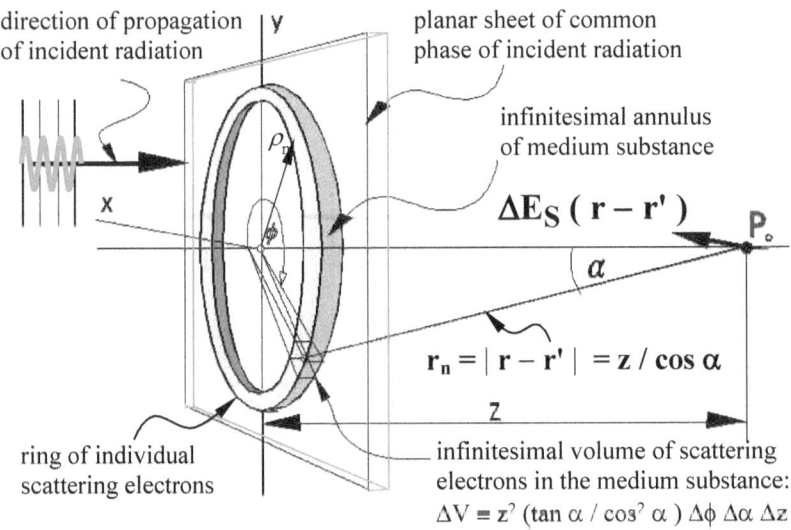

Figure 19.5: Geometrical considerations for integrating the effects of forward scattering

Notice that the annulus corresponds to points near an x,y plane that are all approximately equidistant from P_o. Successive annuli of synchronized scattering emanations from embedded electrons whose distances from P_o incrementally increase by $\frac{1}{2}\lambda$ will alternate between interfering positively and negatively with regard to radiation coming from electrons at the center. The distance for each such annulus can be represented as, $r_n = z + \frac{1}{2} n \lambda$. These zones are effectively *Fresnel zones*. The radii ρ_i of each annulus measured from the center satisfies:

$$\rho^2_n + z^2 = (z + \tfrac{1}{2} n \lambda)^2$$

So that,

$$\rho^2_n = n z \lambda + (\tfrac{1}{2} n \pi \lambda)^2 \cong n z \lambda,$$

if $z \gg \lambda$ as it certainly will be.

Clearly we are assuming single-scattering over the length z. Then if we consider the area within each zone, we obtain:

$$A_n = \pi \, (\, \rho^2_n - \rho^2_{n-1} \,) = \pi \, z \, \lambda$$

So that the area of each zone is the same as for any other and, therefore, the intensity contributions from each zone will be the same even though they are increasingly narrower as α increases with wave phase.

Since the phase varies even within each 'half-period zone', the resultant field strength from each is $2/\pi$ what it would be if the phase were the same throughout the zone. The phase of the contributions of successive annular zones alternate with amplitude slowly decreasing as α increases, so that there is a series which, when added together, sums to $1/\pi$ times the effect that would result from only the first zone if the phase were to have been identical throughout. Thus, by using this factor, we need only include in our analyses the first zone for which,

$$\tan \alpha = \, (\, \lambda \, / \, z \,)^{\frac{1}{2}}$$

This procedure greatly simplifies integrating the effects of scattering.

We have yet to address a means to simplify linear integration all the way from the source to the point of observation. We do that now.

the concept of 'extinction' in a scattering medium

Statements of what is called the *extinction theorem* (Born and Wolf, 1980; Wolf, 1971, etc.) all aver that in an extensive medium, the incident radiation will be totally replaced by forward scattered radiation after having propagated to a distance within the medium, δ_0, known as the *extinction interval*. This distance at sea level in our atmosphere, for example, is on the order of only a millimeter or so. In the intergalactic medium, in accordance with the sparse electron density of this medium as presented earlier, it can be many thousands of light years.

Throughout any scattering medium, in any case, the radiating fields are continually being replaced. The question is, how long and how far does that process require? Replacement fields, following a first replacement, will propagate at the velocity dictated by the index of refraction, **n**, of the medium which thereafter completely characterizes propagation through the medium. Thus, comprehensive tests of the *Second Postulate* of special relativity, as for example Brecher (1968), have had to take this process into account in confirming that postulate to a high degree of accuracy with pulsar data obtained from within our own galaxy and more distant globular clusters. Brecher performed his analyses using the extinction distance formula whose derivation will be developed here. If this concept is already understood by the reader, there are yet aspects of the methodology that need to be understood.

We will use the extinction distance, δ_o, after which incident radiation is replaced by *forward-scattered* radiation, to further restrict a volume enclosed by a conceptual horn shown in figure 19.6 that we will refer to as the '*coherency domain*' of the point P_o. Scattered radiation emanating from charges within the surface of this horn can legitimately be considered as limiting the integration calculation that determines a total composite forward-scattered field $\mathbf{E}_{net}(\mathbf{r})$ at P_o, thus affecting the extinction process. The horn is defined as being elongated just enough so that there will be replacement of the incident field due to the similar but oppositely directed forward scattered field. That field is thus completely out of phase with the incident wave at the point, P_o.

$$\mathbf{E}_{net}(\mathbf{r}) = \mathbf{E}_i(\mathbf{r}) + \mathbf{E}_S(\mathbf{r}) = -\mathbf{E}_i(\mathbf{r})$$

To effect eventual replacement of the incident radiation, the scattering field must eventually overpower the incident field, $\mathbf{E}_i(\mathbf{r})$:

$$\mathbf{E}_S(\mathbf{r}) = -2\,\mathbf{E}_i(\mathbf{r})$$

where $\mathbf{E}_S(\mathbf{r})$ is the resultant of scattered fields derived earlier. It is:

$$\mathbf{E}_S(\mathbf{r}) \equiv (1/\pi) \int_{V'(\text{horn})} \Delta'\,\mathbf{E}_S(\mathbf{r}-\mathbf{r}')$$

where $\mathbf{r}$ is evaluated at P_o and $\mathbf{r}'$ specifies the position of electrons within the infinitesimal volume of integration denoted in figure 19.6 as $\Delta V'$. By substitutions like those performed for individual electrons above, a scattered electric field strength, $\Delta'\mathbf{E}_S(\mathbf{r}-\mathbf{r}')$ can be obtained for each infinitesimal segment of volume in the coherency horn as follows:

$$\Delta'\mathbf{E}_S(\mathbf{r}-\mathbf{r}') = \Delta V'\,(\omega^2/c^2)\,(n-1)\,\mathbf{E}_i\,/\,2\pi\,|\,\mathbf{r}-\mathbf{r}'\,|$$

$$\Delta'\mathbf{E}_S(\phi,\alpha,z) \cong \Delta V'(\phi,\alpha,z)\,(\,2\,/\,\lambda^2\,)\,|Re(n-1)|\,\mathbf{E}_i\,\cos\alpha\,/\,z$$

The net scattered field at any point within the medium can then be obtained as the integral all these infinitesimal fields.

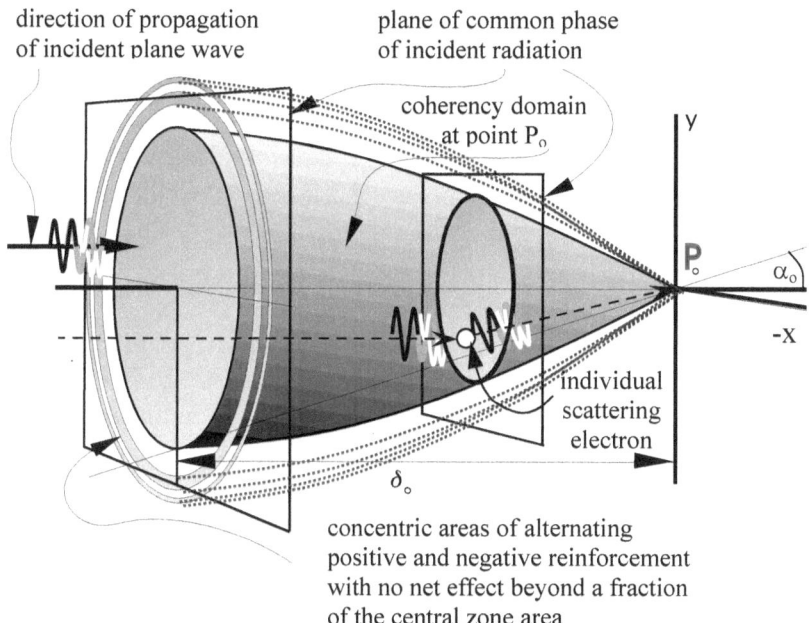

Figure 19.6: Coherency domain in a scattering medium

integrating coherent scattering effects

Clearly the net effect of incident radiation includes the contributions due to polarization of the medium (scattered electromagnetic fields induced by the incident radiation) resulting from all intermediate electrons in the medium. This contribution is determined in large part by phase considerations of the various components received from each scattering electron throughout the medium as discussed above. Ultimately coherent reinforcement of all of the various forward-scattered fields over the total volume of integration results in sufficient scattered field intensity to effect the replacement (or what is called *extinction*) of the incident radiation by the virtually identical scattered fields that have an opposing phase.

Geometrical considerations are typical of integration schemes to be employed throughout this treatise. It is, of course, simplified by considerations just discussed whereby the angular bounds of integration can be reduced to include only the first half-period zone as elaborated in figure 19.6. Importantly, although the result of the integration we will obtain is well known, understanding the constraints of integration illustrated in these figures is key to understanding the alterations that will be required to accommodate for the similar, *although significantly different*, process in a hot plasma in the next chapter.

By restricting the angle as also described above, one can proceed using only the first Fresnel zone as if all scattered radiation from within this angle shares a common phase. Thus the overall range of integration will be greatly reduced by the allowed angular restriction identified above as $\alpha = \tan^{-1} [\, \pi \lambda \,]^{\frac{1}{2}}$. Similarly, the length over which integration must take place need only extend from zero to the extinction distance of the medium, δ_0 rather than the all the way to the source of the incident radiation. So that an integration, which must take into account entire planes of synchronized scattering events induced by incident plane wave radiation and further extended to include planes from the point where evaluation takes place to the source of the radiation, is greatly simplified. The envisioned integration procedure obtains an accurate value for net electromagnetic scattering fields experienced at any point P_0.

Note the usual assumption in any scattering analysis, i.e., that the effects of scattered fields from charges closer to the point at which evaluation takes place than several wavelengths will cancel. This typical assumption is certainly realized statistically in the intergalactic medium for which the likelihood of there even being a scattering electron within this radius is zero to a very high degree approximation.

We will perform this integration here to determine an extinction distance for the intergalactic medium, which is essential to determining the cosmological effects of scattering.

determining the extinction distance and base angle

Having established simplifying limits of integration appropriate for the horn illustrated in figures 19.5 and 19.6, with δ_0 defining the extinction interval at which total extinction is accomplished and α_0 the angular limit at the distance δ_0, we will proceed by instantiating the parameters of integration defined in figure 19.4, as follows:

$$dV'(\phi,\alpha,z) = z^2 \, d\phi \, \tan \alpha \; -\frac{\partial}{\partial \alpha} \tan \alpha \, d\alpha \, dz \; = z^2 \, \frac{\tan \alpha}{\cos^2 \alpha} \, d\phi \, d\alpha \, dz$$

$$E_S = [4 \,|Re(\mathbf{n}-1)| \, E_i \, / \, \lambda^2] \int_0^{\delta_0} \int_0^{2\pi} \int_0^{\tan^{-1}(\lambda/z)^{\frac{1}{2}}} (\cos \alpha \, /z) \, z^2 \, \frac{\tan \alpha}{\cos^2 \alpha} \, d\phi \, d\alpha \, dz$$

$$= 4\pi \, |Re(\mathbf{n}-1)| \, \delta_0 \, E_i \, / \, \lambda \, , \text{ for } z \gg \lambda$$

Since we must have $E_S = -2 \, E_i$ to effect replacement, we obtain:

$\delta_o = \lambda / 2\pi \, |Re(\mathbf{n}-1)|$

This is the derivation of the accepted average extinction interval formula. Clearly, a similar analysis applies to any point within any isotropic homogeneous medium. Brecher (1968) and others have assumed this formula in their analyses of extinction in intra- and near extra-galactic plasmas.

To assess the value of the derived extinction distance in terms of properties of any medium, we must substitute the parametric expression for $Re(\mathbf{n}-1)$ obtained earlier. It was:

$$Re(\mathbf{n} - 1) = 2\pi \, (\, \rho_e \, e^2/m_e) \, (\omega_o^2 - \omega^2 \,) \, / \, ((\omega_o^2 - \omega^2 \,)^2 + \, (\, \gamma \, \omega \,)^2 \,)$$

if $\gamma \ll \omega$ and $\omega_o \neq \omega$ we obtain for the leading factor when $\omega \gg \omega_o$:

$$m_e \, \lambda \, / \, (4\pi^2 \, \rho_e \, e^2 \,) \cong 1.00 \times 10^{-10} \; \lambda \, / \, \rho_e$$

So that we obtain for the extinction interval:

$$\delta_o \; \cong \; 3.5 \times 10^{12} \, / \, \lambda \, \rho_e$$

If we apply this formula to optical wavelengths for which a wavelength of 3,500 Angstroms is representative we obtain:

$$\delta_o \; = 7.12 \times 10^{17} / \rho_e$$

Let us look at the relationship of electron density in a thermodynamic gas: The volume of one mole of a gas is:

$$V_m = \tfrac{2}{3} \, N \, R \, T \, / \, P$$

One atmosphere of pressure is 10^6 dyne/cm^2 so the volume containing one mole of a gas at this pressure and at room temperature (≈ 300 K) is:

$$V_m = 0.667 \times 300 \times 8.3143 \times 10^7 \times 10^{-6} \; = 1.661 \times 10^4 \; \text{cm}^3$$

We know there is Avogadro's number of molecules in each mole of a gas. But it depends upon the molecular structure as to how many optically active electrons are included per mol. There are typically many electrons per molecule, many affected by incident electric fields.

249

Let us take as example the helium molecule for which there are 2 total electrons per molecule. One mole of helium at standard pressure and temperature contains the following number of electrons:

$$\rho_e = 2 \times N_A / V_m = 2 \times 6.025 \times 10^{23} / 1.661 \times 10^4 = 5.804 \times 10^{20} \text{ cm}^{-3}$$

So under these conditions, we obtain for the extinction interval in helium gas,

$$\delta_0 = 7.12 \times 10^{17} / \rho_e = 9.816 \times 10^{-4} \text{ cm}$$

Inner electrons in larger molecules are shielded to varying degrees and each will have a unique resonant frequency that keeps there from being much of a multiplicative effect. Thus, the extinction interval for visible light may be a distance of as little as 10^{-3} cm at atmospheric pressure and temperature or as much as thousands of light years in traveling through intergalactic plasma with respective expected values of electron density. But the same basic formulas apply.

Notice also that the values of the principal angle from a point at the tip of the horn to the base of the coherency domain at this extinction distance is the following:

$$\alpha_0 = \tan^{-1} (\lambda / \delta_0)^{\frac{1}{2}} \cong \tan^{-1} (0.597) \cong 30 \text{ degrees}$$

So that for visible light the conceptual 'horn' depicted in figure 19.6 would be fairly broad, but extremely tiny.

If we look at the total number of scattering electrons N_{horn} involved in the extinction process within a single coherency domain (refer again to figure 19.6), we must first assess its volume V_{horn} for which we obtain:

$$V_{horn} = \pi \sin^2 30^\circ \int_0^{\delta_0} \delta^2 \, d\delta = \frac{1}{3} \times 0.26 \pi \delta_0^3 \cong 3 \times 10^{-13}$$

For this case of visible light, the number of electrons involved in effecting extinction in such a thermodynamic gas will be:

$$N_{horn} \approx 2 \times 10^8 \text{ electrons.}$$

characterizing the limitation on spectral invariance

However strange the scattering processes might seem to those not familiar with them, the replacement of radiation by extinction in the

250

forward scattering has not seemed in itself to be capable of altering the characteristics of the incident radiation. Investigation into the generally observed invariance of the spectra of electromagnetic radiation after having propagated to a considerable distance from its source through a scattering medium by Wolf and others (1971, 1972) lead Wolf (1986) to his discovery of a general scaling law. That law specifies the conditions under which this traditional cornerstone of spectroscopy remains valid.

Understanding this law allowed Wolf (1987 and 1989), James et al. (1990), and others to discover source correlations, which by virtue of violations to these specified conditions generate radiation whose spectrum differs from that which was originally emitted. Using the well-known analogy between the processes of emission of incident and scattering of secondary electromagnetic fields, Wolf et al. (1989) were able to show that the predicted modes of spectral variation of electromagnetic fields in spatially distributed primary sources apply also for randomly distributed scattering media whose constituents act as secondary sources of radiation more or less as we showed in earlier sections of this chapter. In fact regular arrays of secondary scatterers shown in figure 19.2 in reference to Huygens principle are unnecessary.

In other investigations they were able to artificially construct distributed secondary sources of electromagnetic radiation that actually exhibited exceptions to this rule. However, single-scattered emanations used in their investigations do not support forward scattering through multiple extinctions. The degree of alteration of the spectra in these experiments was produced by the introduction of a commensurably appreciable *divergent* scattering angle. Their results bear consideration in obtaining the conclusions we have with regard to their being angular *convergence* rather than *divergence* of scattered radiation in extremely high temperature thermal media.

There is a simple analogy between distributed primary sources of radiation emissions and electron scattering sources associated with the constituents of a scattering medium, by which one can derive the analogous forward scattering 'scaling law' for the invariance of spectra propagated to what is referred to as the 'far zone' through a scattering medium. The isomorphism of the analogy to *primary* sources is so complete that other than the differences we will identify as associated with the relativistic motions of high temperatures, the same conditions for invariance result. This might (prematurely as it turns out) suggest that radiation should also penetrate a thermal medium without experiencing the slightest spectral variation. But that is not the case.

This premise of spectral invariance has been generally presumed with demonstrations of the validity of the presumption readily available. For example, Fraunhofer diffraction patterns of lines in the solar spectra correlate extremely closely with emission spectra obtained in the laboratory for the associated elements other than for the minor redshifts apparent in the limb of the sun for which various explanations have been proposed (and for which we will add one more with more validity). This is true also of other stars within our galaxy and neighboring galaxies with any differences in the spectra typically being accounted for directly as Doppler effects due to peculiar motions of the sources.

Thus, ultimately Wolf's law specifies the coherency conditions between electromagnetic fields emanating from constituents such as e_i and e_j of a distributed source as illustrated in the previous figures 19.1 through 19.3. In the analyses performed by Wolf et al. (1989) of scattering media by analogy with primary distributed sources, they considered only a single extinction/re-radiation process (pretty much as has been done here). Predictions of Doppler-like shifts in spectra in Wolf's later investigations (1987 and 1989) depended on observations of scattered light that was not aligned with the direction of propagation of the incident radiation. Therefore, these predictions pertain only to single-scattered radiation and do not support an accumulation of the effects of forward scattering involving the repeated extinction processes over considerable distances. It might seem, therefore, that their predicted null result in the 'forward' direction after a single extinction is significant for forward scattering situations pertinent to all thermal media we will consider. In as much as spectra remained unchanged after a single extinction in this direction, there might seem to be no new phenomenon introduced by conjoining multiple extinctions that could produce a change after many.

20: The Exception to Spectral Invariance in Thermal Media

"The most obvious alternative, a tired-light model, is not mentioned in many reviews of cosmology, given short shrift in others... Not least of the reasons for the widespread acceptance of the expansion hypothesis is the lack of a reasonable alternative basis for redshift... "[22]

In the previous chapters we investigated the effects of scattering applicable to virtually any medium; these effects certainly apply to thermal media in general although not without caveat to *extremely hot* media. All these former investigations employed the usual practice of excluding from the formal assumptions of their analyses all physical phenomena whose effects are negligible if their inclusion would only obscure an otherwise straight forward conclusion. However, an effect that remains completely negligible after one occurrence might still accumulate to a measurable effect after many – particularly after billions. This is especially true if the value of that *negligible* amount could be shown to always possess the same arithmetic sign.

Previous investigations also very explicitly excluded considerations of the relativistic motions of the individual scattering electrons. This is a limitation that was valid in supporting their restricted results. Again however, this is an inadequate assumption in cases involving extremely high temperatures – particularly for high temperature plasma electrons.

---

[22] Geller and Peebles (1972), pp. 1-5.

These conditions of spectral invariance are not strictly satisfied by even a *primary* thermal source because of individual radial Doppler shifts associated with radial motions of the constituent molecules. It has been repeatedly demonstrated that radiation from such a source results in the well-known spectral line broadening. This occurs because the velocities of the individual molecular or atomic sources of the radiation cause Doppler shifting of the wavelength of the spectral lines, some of which will be shifted to longer wavelengths and some to shorter. But to a very high degree of accuracy the mean of these wavelengths will not have been altered from that of a stationary source. Transverse Doppler shifts will also occur, of course, but since that effect is much smaller than radial Doppler, the effect will be negligible. Since transverse Doppler will be of second order in v/c it will therefore typically be insignificant relative to the radial Doppler effects which will be first order in v/c, where v is the velocity of the source of a particular photon. But line broadening is due to motions of individual sources within a distributed source and is therefore due to the motions of the *primary*, rather than *secondary*, sources of radiation.

So, on all these counts spectral invariance is an issue worth further investigation. The avenues of interest in this regard are the effects of *extremely high temperatures* and those associated specifically with the motions of the *secondary* sources across regions of an intermediate medium where *considerable forward scattering extinction events* occur.

sources of minor variations in spectra

The effects of radially oriented thermal motions of the *secondary* sources toward or away from the direction of propagation of the incident radiation, but with a zero mean velocity distribution, will be reduced to Doppler effects that are higher than first order in v/c after repeated photon replacement. Although the wavelength of scattered radiation from a single electron may be lengthened, on average the next instance of scattering will shorten the wavelength by a similar amount, etc.. It will not precisely cancel out, however, because even an identical radial velocity in the opposite direction will produce a wavelength change that will not exactly cancel out the first change. But canceling plus and minus signs due to successive radial velocities leaves second and higher order terms as follows:

$$\lambda_1 = \lambda_0 \, \gamma \, (\, 1 \pm v/c \,)$$

$$\Delta\lambda_R = \lambda_1 - \lambda_0 = \lambda_0 \, \gamma \, (\, 1 \pm v/c \,) - 1 \cong \lambda_0 \, (\pm v/c + \tfrac{1}{2} \, (v/c)^2 \pm \ldots)$$

The term with the ± sign will have a zero average after many scattering events. Because the likelihood of velocities aligned with the incident radiation direction is small, the total effect would be very small.

As we saw with regard to mediated direct interactions between molecules in previous chapters, there is always a transverse Doppler effect with *any* relative motion of the source of the radiation with respect to an observation. This is because of the gamma factor in the relativistic Doppler equation. This effect occurs also for the scattering electrons along the path of the incident radiation propagation. So when a scattering electron moves at an arbitrary angle θ, rather than strictly radially along the direction of the incident radiation, the 'replicated' scattered radiation wavelength will be altered as follows:

$$\lambda_1 = \lambda_0 \, \gamma \, (\, 1 - v/c \cos \theta \,)$$

This is shown in figure 20.1. The transverse Doppler formula for changes in wavelength associated with motions that are perpendicular to the direction of the incident radiation for which $\theta = \frac{1}{2} \pi$ become:

$$\Delta\lambda_T = \lambda_0 \, \gamma \, - \lambda_0$$

$$= - \lambda_0 \, (\, \gamma \, - 1 \,) \approx \lambda_0 \, (\, \tfrac{1}{2} \, (v/c)^2 \, + \tfrac{1}{4} \, (v/c)^4 + ...)$$

Thus, the scattering electrons produce a net sign independent *second* order effect in v/c on every scattering event throughout an extinction interval. The probability of transverse motions is larger than for radial motions because of the geometrical situation that can easily be seen in figure 20.2. So that when we average the amount of Doppler shift over all possible angles there will in fact be a net redshifting effect.

assessing the cumulative effects on incident radiation

There is a disproportionate amount of lengthening of wavelengths as compared with shortening if the entire range of relative velocity possibilities for scattering electrons is taken into account. Therefore, for a given magnitude of relative velocity applicable in the context of the Maxwell/Boltzmann distribution of random motions, this fact must be included in the analysis. In figure 20.2 the spherical scope of all the possibilities is illustrated. So independent of the value of the magnitude of the velocity |v|, for every angle θ, there are $2\pi \sin\theta$ possibilities associated with this value. What this means is that there are considerably more instances of scattering electrons with velocities

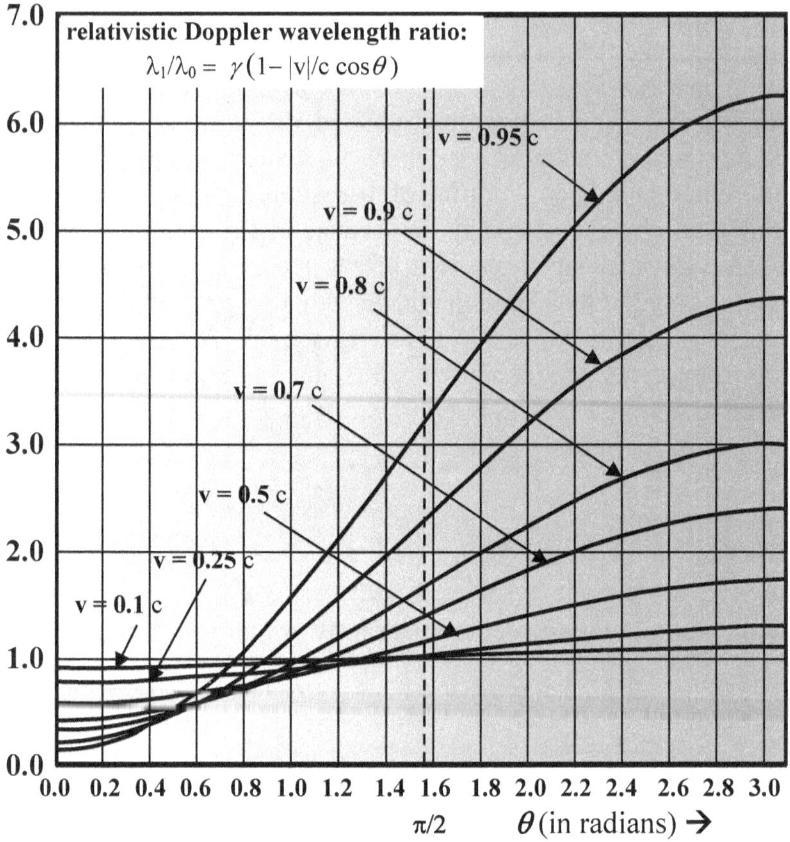

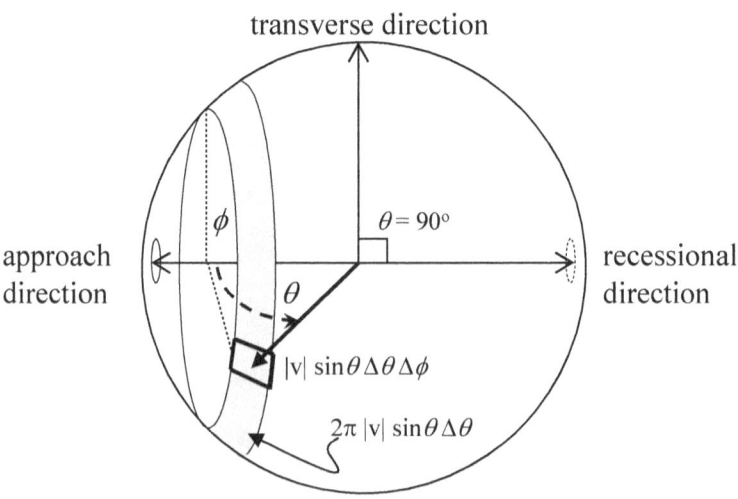

relativistic Doppler wavelength ratio:
$\lambda_1/\lambda_0 = \gamma(1 - |v|/c\cos\theta)$
v = 0.95 c
v = 0.9 c
v = 0.8 c
v = 0.7 c
v = 0.5 c
v = 0.25 c
v = 0.1 c
7.0
6.0
5.0
4.0
3.0
2.0
1.0
0.0
0.0 0.2 0.4 0.6 0.8 1.0 1.2 1.4 1.6 1.8 2.0 2.2 2.4 2.6 2.8 3.0
π/2
θ (in radians) →

Figure 20.1: The net Doppler factor versus angle for various values of v

transverse direction
approach direction
recessional direction
ϕ
$\theta = 90°$
θ
$|v|\sin\theta\Delta\theta\Delta\phi$
$2\pi|v|\sin\theta\Delta\theta$

Figure 20.2: Orientation of electron motions about a propagation direction

in the transverse direction than aligned with the direction of the incident radiation for which there would be cancelation. In the plot of figure 20.1 it is clear that the amount of wavelength lengthening is greatest for recessional velocities. However, these will largely cancel after repeated extinction events as discussed above, but the amount of lengthening that will result for scattering electrons with angles near $\pi/2$ will not cancel. For a given velocity, the average of the wavelength shift will always be positive and dependent only on the velocity of the electrons.

In every conceivable mundane case of forward scattering these wavelength shifts will be negligible since the velocities of molecules will be very insignificant with regard to the speed of light such that the probability density of higher order ratios of v/c will always remain inconsiderable. However, however minor the effect may be, *it is an irreversible loss of energy in the incident radiation that is transferred to the kinetic energy of the medium through which it propagates.* Conservation of energy and momentum are maintained, but entropy increases nonetheless, equalizing the energy in radiation and particulate kinetic energy.

The suspect concept of 'tired light' as envisioned early on by Edwin Hubble as a possibility to account for the redshifted spectra of distant galaxies in the context of cosmology is in fact a reality. There is in fact a dissipation of its energy by scattering. The effect will accumulate with every photon replacement by extinction. Thus, however tiny these relativistic Doppler effects may be in any particular situation, they will accumulate without limit in an extended high temperature, low density medium for which absorption of the originally transmitted (and continually replaced) photon does not occur for a long period of time (distance). Although these transverse Doppler effects are nearly always negligible in any one instance, they are additive and therefore may accumulate to the ultimate significance if the medium is sparse and deep and at a high enough temperature so that the random electron velocities become very large and $(v/c)^2$ becomes more appreciable.

assessing the magnitude of wavelength lengthening

Clearly, the amount that the incident radiation will be redshifted at each extinction interval depends intimately upon the temperature of the medium through which it propagates. Thus the Maxwell/Boltzmann distribution determines what is to be expected of the non-unit terms of the wavelength modification caused by the forward scattering effect.

The root-mean-squared velocity demonstrated to characterize that distribution was, $<v^2>/c^2 = 3\,k\,T\,/\,m\,c^2$. So we can solve for $<v^2>$ in terms of T to obtain the approximate correspondence:

$<v^2> \approx 4.54 \times 10^{-10}$ T

To estimate wavelength change per extinction interval for various values of the temperature of the system we will assume a hydrogen plasma gas here. When the temperature reaches about 10^4 K there is a disassociation of electrons so that the mass to be considered is only that of the electron rather than that of the atom or molecule as a whole, so the wavelength change is larger by the additional factor of 1,835 which is the ratio of the mass of a baryon to the mass of the electron. So to realize an expected electron velocity of 0.5 c, we would require the medium to be at a temperature of T = 1.2×10^{10} K. In figure 20.3 we show the expected single extinction redshift from an electron at various angles to the incident radiation direction for a set of values of temperature of the scattering medium.

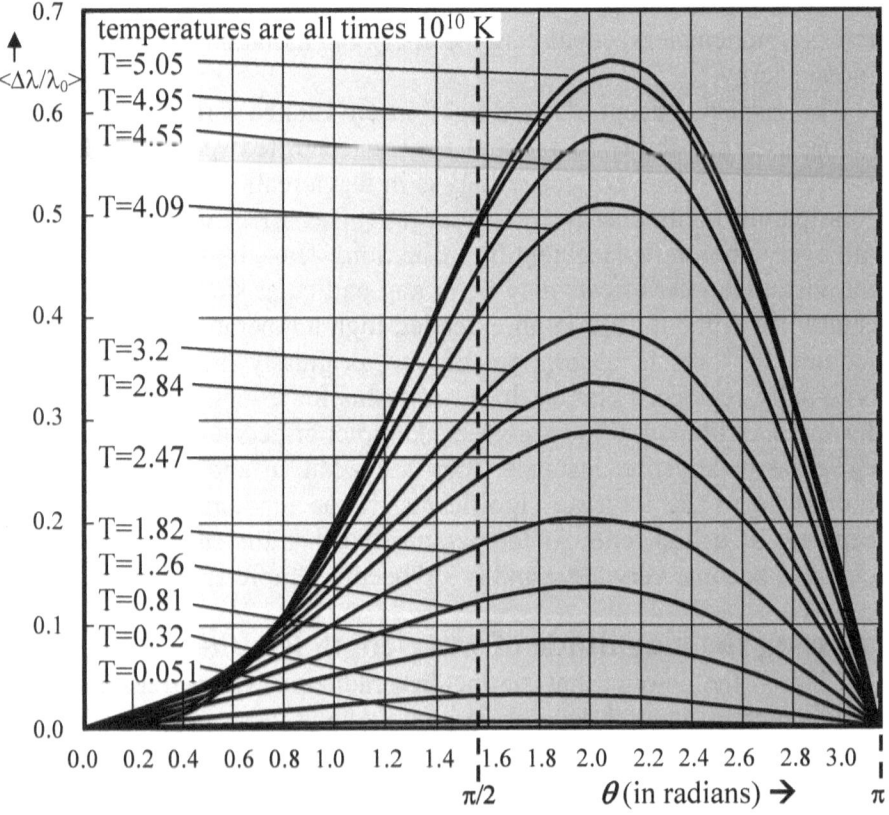

Figure 20.3: The expected single extinction redshift for various scattering electron velocity angles and values of medium temperature, T

Here we consider only the average transverse Doppler shift in wavelength incurred; the results are quite similar to that of the total effect. If we direct our attention to the relationship of this average to medium temperature of the medium we obtain a very essentially linear relationship:

$$< \Delta\lambda_T/ \lambda_0> = 3 \text{ k T} / 2 \text{ m}_e \text{ c}^2 = 2.52 \times 10^{-10} \text{ T}$$

From figures 20.1 and 20.2 it is clear that the weighted average wavelength change per electron per each extinction interval will virtually always be positive, although of course, virtually always negligible when considered by itself.

There will be a distribution of wavelengths of scattered radiation λ about the mean on each extinction. These will naturally be distributed in accordance with the Maxwell/Boltzmann distribution. But as scattering continues, one extinction interval to the next, the constraint of single-scattering of radiation requires that resulting wave functions be within a quarter wavelength in order to constructively interfere with each other. This focuses the radiation about a mean redshift value.

conservation laws applicable to radiation scattering

We have discussed the conservation of energy and momentum at some length in earlier chapters. But those discussions involved aspects of total exchange and transfer of energy and momentum from photons to atoms and/or molecules and vice versa. Scattering obviously does not involve the total transfer of energy and momentum from photons to the scattering electrons – only a portion of that energy and momentum is transferred. To understand the difference and how that translates to the net change in wavelength after each extinction event, consider the diagram in figure 14.3 in the earlier chapter which illustrates the impact of what is called Comption scattering.

Of course Compton scattering per se involves X-ray photons that are being deflected by stationary electrons. Significantly, however, the concept illustrates what is involved in partial transfers of energy and momentum between radiation and electrons in general. To apply the principles to forward scattering of a broader spectrum of radiation as was done by Bonn (2009) one must stress analogies of multiple high speed scattering electrons converging at the locally stationary point at the tip of their respective coherency domains.

Since the velocity of radiation in a vacuum (between scattering events) is independent of relative velocity of the emitter and detector of the radiation, in order for radiation scattered from two electrons with appreciable relative velocity to constructively interfere with one another, they must be affected by incident radiation from an earlier phase plane that is the same distance from each as shown in figure 20.4 (the distance being δ_o) taken again from Bonn (2009).

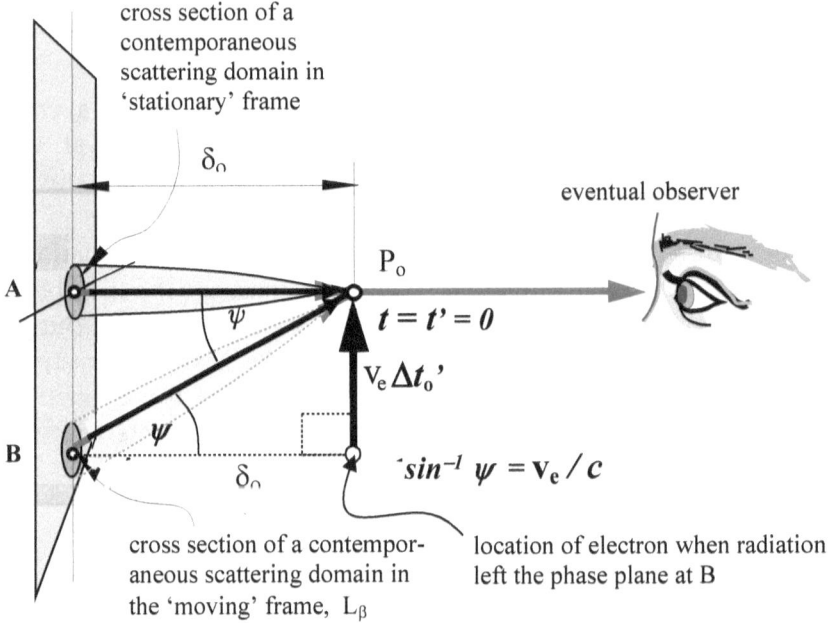

Figure 20.4: Geometrical and temporal relation between scattering events for relatively moving electrons

A 'stationary' electron at P_o could not detect radiation from scattering events from area B detected by the 'moving' electron until hours later than that from his own domain centered at A. There is an aberration of comparative angles, so that ultimately coincident electrons P_o would not detect the same secondary radiation while in coincidence. Scattering from the circular neighborhood about the point B on the planar surface in figure 20.4 will exhibit all the features realized around the centerline of the coherency domain whose base area is about the point A for the 'stationary' electron. So the point is that there will be a separate coherency domain for virtually every scattering electron in a high temperature sparse medium in intergalactic space in particular.

As an electron characterized by the velocity v_e progresses to its coincident position at P_o, various planar surfaces will effectively constitute the scattering surfaces whose effects combine at the base of the horn at t'=0 in his frame of reference. These cross sections which are only aligned in his frame of reference will occupy positions with regard to P_o (in the 'stationary 'frame) as shown in figure 20.5. Altogether these coherency domains constitute a single angled domain in the 'stationary' frame of reference suggested in figure 20.4.

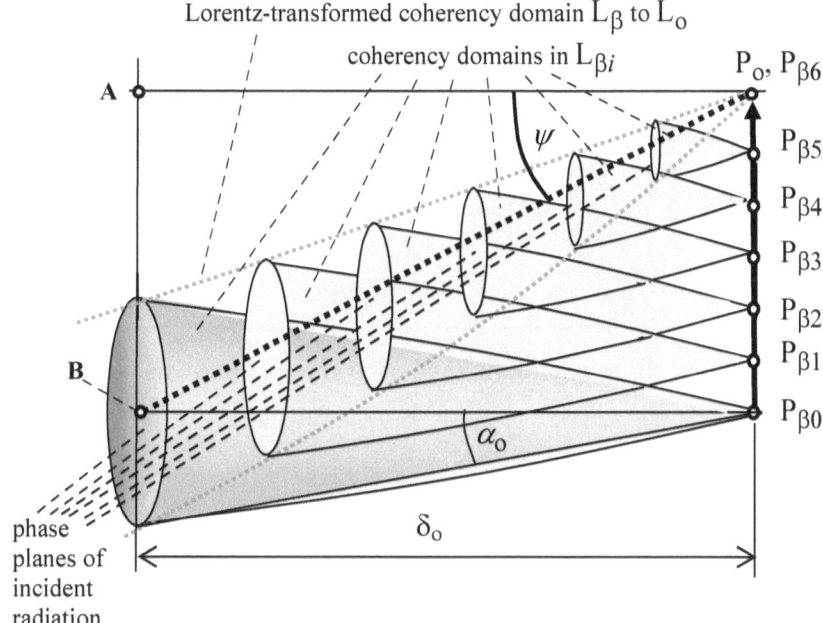

Figure 20.5: Geometrical relationship between the coherency domain in L_o and its Lorentz-transformed counterpart in L_β

Notice that there is no pretense that the scattering events which are detected are the very same in both reference frames; clearly they typically will not be. But – and this is an important point – although there are Lorentz time differences to the respective points on the plane when viewed in a *single* frame of reference, these differences are entirely accounted for by differences in propagation times in the two *separate* frames if the same radiation were to be assumed as being detected while in coincidence. In other words the time difference is *not* in the emission clock times from areas A and B in figure 20.4 and 20.5, nor, therefore, in the phase of the incident radiation at the time of those scattering events that will so variously be detected.

Thus, in applying the conservation concept employed by Compton, there is a deflection of a photon and a velocity difference before and after radiation proceeds from the point P_o (an ensemble of secondary emission sources, i. e., scattering electrons) with a generally forward direction that combines their total effect. So it is from relatively moving secondary sources with radiation coming from a locally stationary area. To solve the conservation equations that were illustrated in figure 14.3, we begin with:

BEFORE		AFTER
	energy	
$h\,c\,/\,\lambda_{\text{initial}} + \gamma_1\,m_e\,c^2$	$=$	$h\,c\,/\,\lambda_{\text{final}} + \gamma_2\,m_e\,c^2$
	momentum	
$h\,/\,\lambda_{\text{initial}}$	$=$	$h\,/\,\lambda_{\text{final}}\cos\psi + \gamma_2\,m_e\,v_2\sin\theta$
$\gamma_1\,m_e\,v_1$	$=$	$\gamma_2\,m_e\,v_2\cos\theta - h\,/\,\lambda_{\text{final}}\sin\psi$

Solving these three equations involving three unknowns, we obtain as Compton did:

$$\lambda_{\text{final}} - \lambda_{\text{initial}} = \lambda_C\,(1 - \cos\psi),$$

where,

$$\lambda_C = h\,/\,m_e\,c$$

is known as the Compton wavelength; it's value is approximately equal to 0.02425 Angstroms = 2.425 x 10^{-10} cm. And we can manipulate the final factor in the wavelength change as follows:

$$(1 - \cos\psi) = (1 - \cos^2\psi)\,/\,(1 + \cos\psi) = \sin^2\psi\,/\,(1 + \cos\psi)$$

Thus, for small deflection angles (which will always be the case) we have that:

$$\lambda_{\text{final}} - \lambda_{\text{initial}} = \tfrac{1}{2}\,\lambda_C\sin^2\psi$$

Ultimately, of course,

$$< \sin^2\psi > \; = \; <v^2>/c^2 = 3\,k\,T_e\,/\,m_e\,c^2$$

And we arrive finally at:

$$\lambda_{final} - \lambda_{initial} = 3 \text{ k h } T_e / 2 \text{ m}_e^2 \text{ c}^3$$

$$= 6.14 \times 10^{-20} T_e \text{ cm}$$

A truly tiny amount of wavelength lengthening at each extinction as well as an insignificant increase in scattering electron velocity.

How does wavelength lengthening relate to redshift?

Certainly this effect does not alter the wavelength of the replaced photon by an appreciable amount after a single extinction interval. But it *does* lengthen it by that miniscule amount and significantly it will do it again on the *next* interval, and on the next, etc.. The effect is cumulative and irreversible. But it must be acknowledged that this lengthening of the wavelength of radiation propagating through a thermodynamic medium does not directly equate to the 'cosmological redshift' attributed to recessional velocities of distant sources.

Redshift is without exception defined as:

$$Z = (\lambda - \lambda_o) / \lambda_o$$

In other words, the redshift Z of the observed radiation is dependent upon the wavelength of the observed radiation λ as well as that of what was originally emitted, λ_o. So that over a single extinction interval we would have:

$$\lambda / \lambda_o = Z_{\delta o} + 1$$

The redshift over one extinction interval δ_o is the smallest increment of redshift that can be realized. So in intergalactic plasma, should we choose to account for wavelength change in that way, we would have:

$$\Delta(Z + 1)_{\delta o} = \Delta\lambda_{\delta o} / \lambda_o = 6.14 \times 10^{-20} T_e / \lambda_o$$

At each stage of the extinction process we have,

$$\lambda_n = \lambda_{n-1} + 6.14 \times 10^{-20} T_e$$

$$= \lambda_{n-2} + 2 \times 6.14 \times 10^{-20} T_e$$

$$\vdots$$

$$= \lambda_o + n \times 6.14 \times 10^{-20} T_e$$

Thus, if we assume a homogeneous medium with essentially the same ambient temperature T_e throughout, we obtain:

$$Z_n = n \times 6.14 \times 10^{-20} \, T_e / \lambda_o$$

This linear relationship is plotted in figure 20.6 below. The wavelength dependence illustrated in the curves plotted in figure 20.6 clearly distinguishes this from a Doppler redshift, each wavelength demanding its own plot. But one could not expect a functionality that was independent of wavelength for redshift versus number of extinction intervals – as against distance – through intergalactic regions, for example. This is because propagation distance between extinction intervals is also wavelength dependent as we will show.

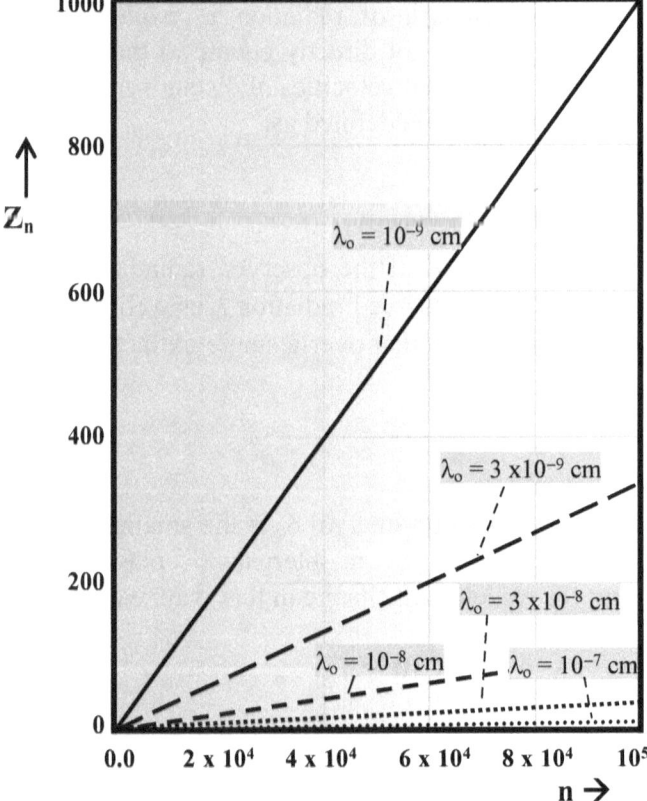

Figure 20.6: Form of relationship for 'redshift' versus number of extinctions

This is certainly not the proportionality of redshift with recessional velocity of the source of radiation or distance that was cited by Hubble and would become a defacto 'cause' of the redshift of distant galaxies.

an alternative basis for the distance-redshift relationship

The effect we have just discussed is not immediately distance-related in the way Hubble had hoped with regard to the redshifted spectra from distant galaxies. Certainly it would increase with distance, but just adding redshift increments is not what is observed. Nonetheless, since it occurs at every extinction interval and does, therefore, increase with distance, the precise nature of the redshift-distance relationship must be investigated based on the parameters that occur in the extinction interval δ_0 calculation. In the previous chapter it was demonstrated that to effect photon replacement, we must have that:

$$\delta_0 = \lambda / 2\pi \, |Re(\mathbf{n}-1)|$$

On investigating the value of the real part of the index of refraction in terms of properties of the medium, we obtained the following:

$$Re(\mathbf{n} - 1) = 2\pi \, (\rho_e \, e^2/m_e) \, (\omega_0^2 - \omega^2) / ((\omega_0^2 - \omega^2)^2 + (\gamma \omega)^2)$$

Substituting the values for the mass and charge of a plasma electron in the inverse of the leading factor, we obtain:

$$m_e / (2\pi \, \rho_e \, e^2) \cong 6.26 \times 10^{-10} / \rho_e$$

Remembering that $\omega = 2\pi c / \lambda$ and employing the usual conditions that, $\gamma \gg \lambda / 2 \pi c$, and $\lambda \ll \lambda_0$, where λ_0 is the resonant wavelength of the medium, which for hydrogenous plasma is extremely long, we obtain as a good approximation for a final factor, ω^2. Thus we have:

$$\delta_0 \quad \cong (\lambda / 2\pi) \, (m_e / (2\pi \, \rho_e \, e^2)) \, (2\pi c / \lambda)^2$$

$$= m_e \, c^2 / e^2 \, \lambda \, \rho_e$$

$$= 3.56 \times 10^{12} / \lambda \, \rho_e$$

If we apply this formula to incident radiation in optical wavelengths for which a wavelength of 5,000 Angstroms is representative, we obtain a value of $\delta_{ov} = 7.12 \times 10^{17} / \rho_e$ cm. In intergalactic plasma this is thousands of light years. Therefore, the incurred increase in redshift *per centimeter* throughout the propagation distance can be calculated as follows:

$$Z_{1cm} \quad = \Delta(Z+1)_{\delta o} / \delta_o$$

$$= [\, 6.14 \times 10^{-20} \, T_e / \lambda_o \,] / [3.56 \times 10^{12} / \lambda \, \rho_e]$$

$$= 1.725 \times 10^{-32} [\, \lambda_i / \lambda_{i-1} \,] \, T_e \, \rho_e$$

The expected redshift per centimeter caused by scattering electrons is a truly miniscule amount in virtually any mundane medium. But as discussed extensively by one of the authors,[23] the phenomenon *does* occur and there are situations for which (in addition to another interesting instance of irreversibility) it makes a phenomenally observable difference. In point of fact it is in the realm of cosmology where it makes a most tremendous difference.

We demonstrated above, however, that for light propagated through many extinction intervals, the lengths of the intervals tend to change significantly due to wavelength lengthening. Therefore in determining the composite effect of traveling an appreciable distance, we must take this into account. The quasi unit redshift Z_{1cm} defined above only assessed the change in redshift per centimeter occurring over the length of a single extinction interval. One can not assume that it would remain unchanged over all intervals as light propagates because of the wavelength change along its path. When extremely long transmissions occur in thermal media, we must accumulate effects over the entire distance. We must take into account that wavelength functionality as extinction interval length continues to change. So the total distance in centimeters must be determined as:

$$r(n) = \sum_{i=0}^{n} \delta_{oi} \cong [\, 3.56 \times 10^{12} / \rho_e \,] \sum_{i=0}^{n} 1 / \lambda_i$$

Thus, if we assume a homogeneous medium with essentially the same ambient temperature T_e throughout, we obtain:

$$\lambda_i \quad = \lambda_{emitted} \, (\, 1 + i \times 6.14 \times 10^{-20} \, T_e / \lambda_{emitted})$$

$$= \lambda_{emitted} \, (Z_i + 1\,)$$

The propagation distance of incident radiation thus becomes a function of number of intervening extinction intervals n as follows:

---

[23] Bonn, *Cosmological Effects of Scattering in the Intergalactic Medium*, 2011.

$$r_n = [3.56 \times 10^{12} / \rho_e \lambda_{emitted}] \sum_{i=0}^{n} 1 / (1 + i\, 6.14 \times 10^{-20}\, T_e / \lambda_{emitted})$$

This propagation distance equation depends upon the number of extinction intervals as illustrated in figure 20.7 taken from Bonn.

When the number of extinction intervals is small, this distance is directly proportional to the number of extinction intervals as follows:

$$r_n \cong n \times 3.56 \times 10^{12} / \rho_e \lambda_{emitted}$$

In which limiting case, we have the distance-redshift relation:

$$r_n \cong [5.80 \times 10^{31} / T_e \rho_e] Z_n$$

The electron density of intergalactic space is extremely low, i.e., $10^{-5} \text{ cm}^{-3} > \rho_e > 10^{-7} \text{ cm}^{-3}$ and the average temperature is extremely high, i.e., $10^3 \text{ K} < T_e < 10^7 \text{ K}$ which supports acknowledged ionization levels. This situation in a scattering medium over such extreme distances in a sparce medium produces what have been otherwise attributed to 'cosmological effects'.

The redshift distance formula conjectured by Hubble was:

$$r = H_o^{-1} Z$$

Here H_o is Hubble's constant, $H_o = 7.14 \times 10^{-29} \text{ cm}^{-1}$. Thus, we have so far obtained a parametric estimate for Hubble's constant, which is:

$$H_o \approx 1.725 \times 10^{-32} [\lambda_i / \lambda_{i-1}] T_e \rho_e$$

In the figure this phase of the relationship is indicated by the curved portions of the curves on the log scale. Clearly, however, it is the logarithmic form of the relationship that applies when propagation is far enough that the number of extinction intervals is much greater than about $4 \times 10^9 / T_e$. The summation can be increasingly approximated by an integral when the number of extinction intervals becomes large so that we obtain the following relationship:

$$r(n) = [5.80 \times 10^{31} / T_e \rho_e] \ln(n)$$

From what we defined above for redshift, we have that:

$$r(n) = [5.80 \times 10^{31} / T_e \rho_e] \ln(Z(n) + 1)$$

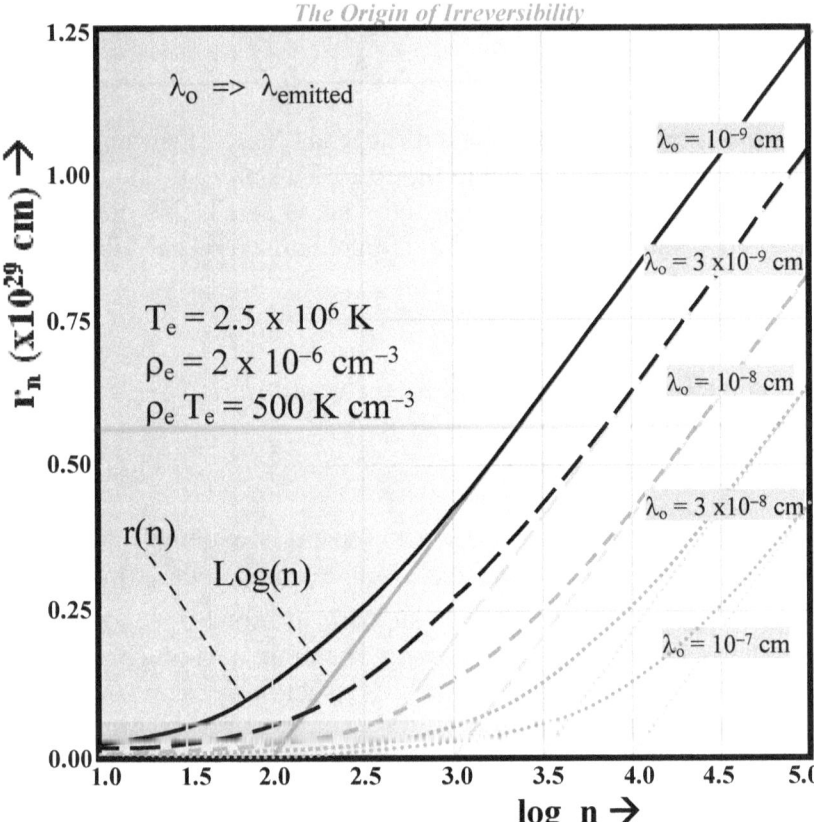

The figure shows curves labeled:

$\lambda_o \Rightarrow \lambda_{emitted}$

$\lambda_o = 10^{-9}$ cm

$\lambda_o = 3 \times 10^{-9}$ cm

$\lambda_o = 10^{-8}$ cm

$\lambda_o = 3 \times 10^{-8}$ cm

$\lambda_o = 10^{-7}$ cm

$T_e = 2.5 \times 10^6$ K

$\rho_e = 2 \times 10^{-6}$ cm^{-3}

$\rho_e \, T_e = 500$ K cm^{-3}

r(n)

Log(n)

y-axis: r_n ($\times 10^{29}$ cm) $\rightarrow$ with values 0.00, 0.25, 0.50, 0.75, 1.00, 1.25

x-axis: log n $\rightarrow$ with values 1.0, 1.5, 2.0, 2.5, 3.0, 3.5, 4.0, 4.5, 5.0

Figure 20.7: The logarithmic form of the relationship between distance and number of extinction intervals

There is clearly an observable distinction to be made at low redshifts, where a fixed relationship between the logarithm of Z(n) and r(n) will not yet have been established. Radiation must travel an appreciable distance as shown in figures 20.7 and 20.8 before redshift becomes an accurate predictor of distance. Until that distance has been traversed Z(n) is not a reliable indicator of distance.

Because the extinction interval is so long for short wavelength radiation a great distance will be traversed before radiation from cosmological objects characterized by such short wavelength emissions experiences redshifting. So objects of longer wavelength in the vicinity will be more accurate indicators of distance. Since there is no similar distinction with regard to redshifting according to any variant of standard cosmological models, these exceptions to usual distance–redshift curves at short wavelengths could provide a means of falsifying cosmological models.

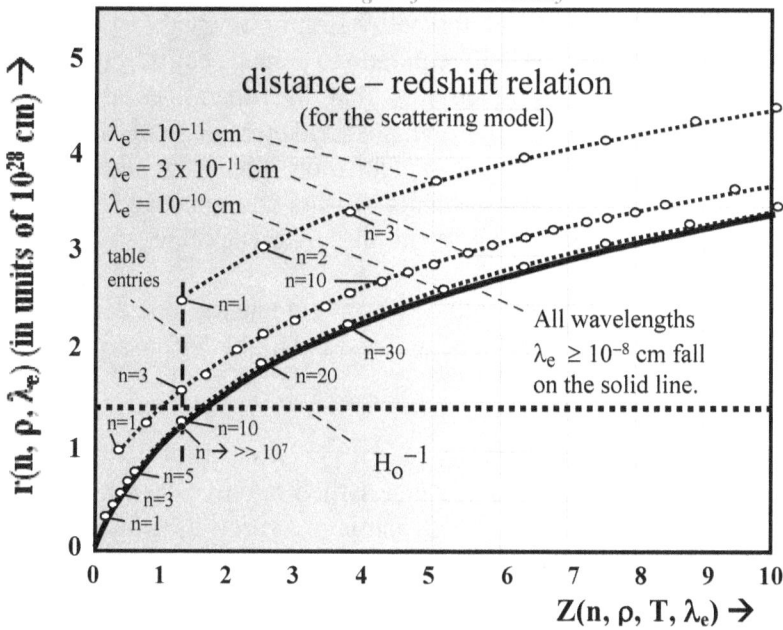

**Figure 20.8: The emergence of a distance versus redshift relation
independent of wavelength over a broad range**

In any case the form of the relationship that replaces the linear approximation at extreme distances is such that the redshift-distance relationship is logarithmic in one direction and exponential in the other:

$$r(Z) = H_o^{-1} \ln (Z(r) + 1) \quad \rightarrow \quad Z(r) + 1 = e^{H_o\,r},$$

Certainly the numbers associated with the changes in wavelength due to forward scattering are not why sunsets are red or other earth-bound wavelength-altering phenomena occur. These effects – to the extent that they are observable – come to play exclusively in the arena of cosmology. But the universe itself is a thermodynamic system and for this largest of all systems, this is how irreversibility plays out.

ramifications of the scattering explanation of redshift

Naturally scattering as the causal basis of cosmological redshift has many ramifications even beyond the obvious difficulties of getting it accepted by a community that is thoroughly enamored with a totally different explanation.

What might be perceived as an objection to the explanation is that the universe involves much more than an intergalactic medium and thus one might argue that it is not just a stable thermodynamic system.

These objections have been addressed by Bonn elsewhere, so suffice it to say that the time period required to reach equilibrium in a thermodynamic system is much less that the operative phenomena occurring in these regions. In the next chapter we will show that although the universe involves much more than an intergalactic medium, the huge occupancy domains of stars and galaxies reduce the effects of 'local' objects relative to the overwhelming vastness of intergalactic regions.

Of more consequence would be a reluctance to accept the parametric expression that must be equated to Hubble's constant:

$$H_o \leftarrow 1.725 \times 10^{-32} \, T_e \, \rho_e$$

If we assume a homogeneous universe which is only reasonable, then we obtain a value for the average dynamic pressure of the intergalactic medium as follows:

$$< T_e \, \rho_e > = 4.14 \times 10^3 \, K \, cm^{-3}$$

Here the kinetic temperature of the plasma electrons and the density of those electrons determine its value. But in as much as there are major variations of temperature and density depending upon the distance of lines of sight from the centers of galactic clusters, it is the average of the product of dynamic pressure $< T_e \, \rho_e >$ that must equal Hubble's constant. The overall homogeneity of the universe on large scales will guarantee that this average is the same in all direction over long enough propagation distances, which is after all the criterion of whether such phenomena can be considered 'cosmological'. This averaging will in itself have tremendous ramifications on the nature of blackbody radiation associated with such a redshifting medium as we will show in the next chapter.

Significantly, however, the variation for lines of sight that traverse through galaxy clusters will encounter temperatures and densities that are orders of magnitude greater than their averages. This will produce much larger redshifts on galaxies on the far side of a cluster than on the near side. When these vast differences are interpreted in accordance with the virial theorem as recessional velocity differences it exaggerates the presumed mass density of these orbital systems, giving rise to rash justifications for 'dark matter' as contributing to gravitational effects even though there are no luminous material objects to which such vast masses could be assigned. Thus by excluding all other causes of

Doppler frequency shifts than recessional motions, current cosmology has gone somewhat astray by resurrecting Einstein's acknowledged 'greatest error' to account for this phenomena.[24]

Where do radiation energy and momentum losses go?

Any redshift in forward scattered radiation results in a loss of both energy and momentum from the incident radiation. Clearly these losses of energy and momentum must be taken up by the particulate aspects of the thermodynamic system in order to satisfy the conservation laws?

Whatever the initial velocities of the constituent particles, there must be a net increase after the incident radiation has passed. So the sums of the energies and momenta of the constituents in the scattering domain depicted in figure 19.6 will differ as follows:

$$\sum (\gamma_2 - \gamma_1) \, m_o \, c^2 = h \, (\nu_1 - \nu_2)$$

$$\sum (\gamma_2 \, v_2 - \gamma_1 \, v_1) \, m_o \, c = h \, (\nu_1 - \nu_2)$$

These are viable (compatible) equations even though if applied to the individual particles they would not be. If applied to individual transactions we would have the following:

$$(\gamma_2 \, v_2 - \gamma_1 \, v_1) = (\gamma_2 - \gamma_1) \, c$$

$$\gamma_2 \, (1 - v_2/c) = \gamma_1 \, (1 - v_1/c)$$

which are not compatible in as much as they only facilitate a single invalid conclusion that $v_1 = v_2$ with no frequency change at all.

In scattering situations it is the ensemble of particles, not every (or any one) particular that takes up the difference. And, in any case it is the ensemble of particles in the scattering medium that retains the residue of energy lost by scattering phenomena. Even though frequencies are typically much lower than in situations studied by Compton in defining what we know as 'Compton scattering', a similar transfer of energy and momentum between the photon and electrons is involved.

---

[24] For a full accounting of cosmological implications of a scattering redshift, refer to Bonn (2011).

21: The Difference between Foreground and Background Thermal Radiation

"A curious dynamic tension has been arising in the field of cosmology. Some widely held theoretical assumptions are coming into increasing conflict with observational results, and yet those assumptions continue to receive strong support."[25]

"In fact, however, step by step their deep divergences and incoherencies emerge increasingly within the scientific community, but people do not see them until finally the confusion becomes so great that the situation breaks down."[26]

Radiation is an integral aspect of any thermodynamic system. When that system is in equilibrium the kinetic energy of the particulate constituents of the system ineluctably become distributed in accordance with the Maxwell/ Boltzmann distribution. In the same way radiation given off by a system in equilibrium will be distributed in accordance with the Planck distribution of blackbody radiation as shown by Einstein's analyses. *Thermalization* is the process whereby these two representative distributions of energy come to maintain a compatibility of temperature that persists as long as there is equilibrium.

The microwave background radiation is one of the more profound of cosmological facts about our universe. It is a simple enough fact that the radiation observed in every direction as we look out into the universe is blackbody radiation in the microwave domain of the electromagnetic radiation spectrum with an associated temperature of 2.725 K as illustrated in figure 21.1. It has seemed natural to assume that the universe itself must therefore be at that temperature. It isn't.

---

[25] Oldershaw (1990)

[26] Kuhn (1962)

determination of the spectrum of background radiation

In figure 21.1 Far-Infrared Absolute Spectrophotometer (FIRAS) instrumentation data obtained from NASA's Cosmic Background Explorer (COBE) satellite is shown. This data is plotted with the abscissa given in waves per centimeter. It shows a theoretical curve for 2.725 K blackbody radiation.[27]

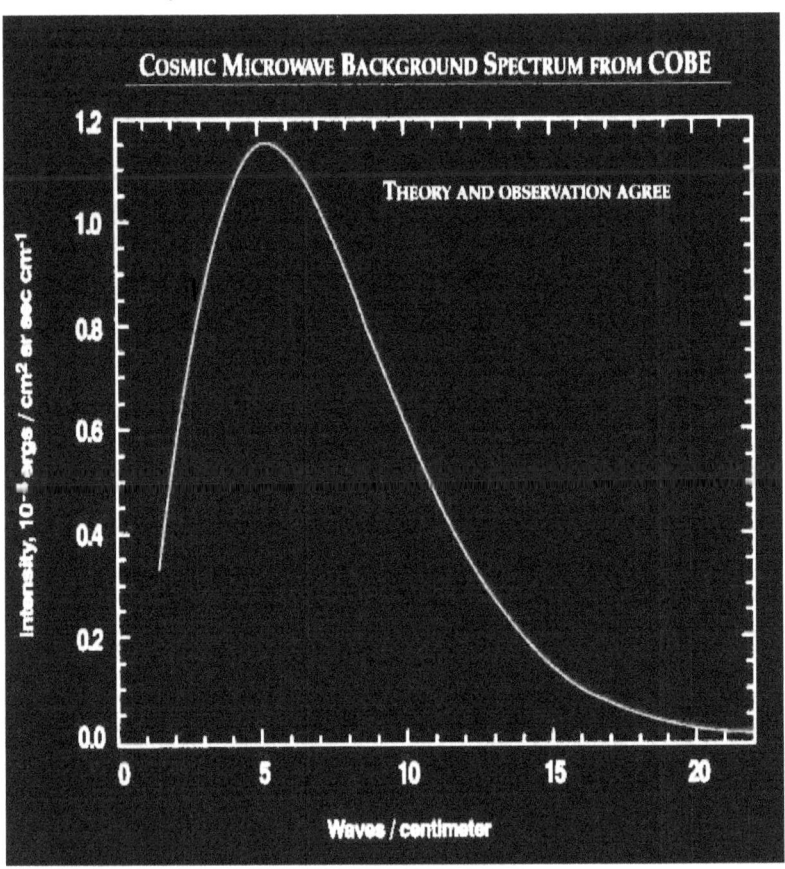

Figure 21.1: Measured microwave background radiation

Notice that the comment "THEORY AND OBSERVATION AGREE" was prominently displayed on the plot in figure 21.1 prior to its dissemination by NASA. Why? The answer can only be that the observed data can very accurately be represented by the theoretical blackbody curve for the temperature 2.725 K. It is accurate, in fact, to within the width of the line used to draw the curve, and therefore, no error bars need to be shown. In figure 21.2 actual tolerances *are* shown

---

[27] Much of the discussion in this chapter is taken from Bonn (2011).

to demonstrate that the data in figure 21.1 is indeed accurate to within the line width of the plotted curve. This data and the figure were taken from Ned Wright (2007) where the error tolerances have been multiplied by 400 and plotted on the NASA data. Clearly the microwave background radiation corresponds to the 2.725 K blackbody spectrum.

However, that is *all* that the widely disseminated comment on the plot legitimately implies. Yet, of course, it was *meant* to imply much more than that. The obvious intention would seem to be that this chart has fully confirmed the aspirations of a generation of cosmologists who had tentatively accepted the veracity of the standard cosmological model before finally embracing it. But, however exceptional the technology that went into obtaining this data, it is *only*, or perhaps more appropriately one should say that it is *superlatively*, just data. It is of course pertinent to assessing *whatever theory* can fully explain it. But it is data, nothing more. It confirms nothing. A theory to explain it need not involve a big bang; the plot is not a snapshot of a big bang.

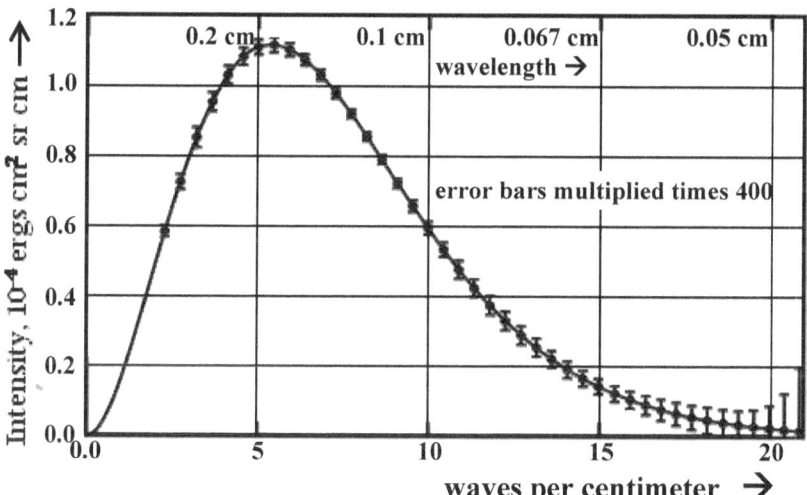

Figure 21.2: FIRAS data with exaggerated error bars included

In any case, the associated temperature of the background radiation is extremely low. Everything we observe in the universe is many orders of magnitude hotter. How can that be? How can blackbody radiation in a system with a high kinetic temperature be cold? This is a thermodynamic issue that stands in direct conflict with all the theories we have learned of blackbody radiation associated with thermodynamic systems generally. And the standard cosmological model explanation skirts the thermodynamic system issues altogether.

A question that no one has seemed willing to ask is, "What would be the radiation spectrum of an extensive medium in which redshifting occurs?" The authors have dared to ask it of those who should know the answer. Their replies suggest that the question is not a legitimate because in their mind there is no such thing as a 'redshifting medium'; scattering is supposed to be associated with spectral invariance. But as we have found, it isn't – not in a hot sparse thermodynamic gas like the intergalactic medium that is orders of magnitude hotter and less dense than anything on earth. So we need to investigate the thermalization process – particularly in an extensive system in equilibrium.

thermal effectiveness of various forms of matter

As we found in chapter 7 with regard to figure 7.5 in particular, the 'blackbody' spectrum is so-denominated because of its association with what was called 'cavity' radiation where everything in the cavity is at the same stable temperature. In thermal media for which there is no 'cavity' per se, the semblance of a cavity wall is produced by the extent and density of the medium itself. Every line of sight must reach an electron in the medium thereby resulting in an outer surface of sources of observed radiation. Otherwise the medium will produce radiation that although exhibiting certain thermal properties, won't be blackbody in form. So how does that concept apply to our universe as a whole?

Importantly, their extreme separations are why the stars and galaxies comprised of them do not contribute much to thermalization that produces thermodynamic balance in the universe. Nor, therefore, do they contribute much toward completing a 'cavity surface' essential to establishing a blackbody distribution of radiation, and most certainly in combination they do not result in the observed background spectrum.

The general topic of the amount of sky cover (or cavity surface area) afforded by various forms of matter, each possessing the same overall amount of mass, is what is of interest here. This is how one determines the distance to a cavity surface analogy in thermalization of energy in an expansive medium with no explicit boundaries.

Let us consider the average sky cover per individual electron in various configurations, since electrons are what are primarily involved in the scattering of usual forms of radiation. Also, electrons have much greater cross sections than do baryons. So they are very much more pertinent to establishment of a semblance of a cavity surface necessary for the associated thermalization process whereby a stable blackbody radiation distribution of energies is produced.

There are on the order of 10^{57} baryons and about that same number of electrons contained in our sun whose cross section is 1.5×10^{22} cm^2.

The average cross section per electron in this configuration is therefore 1.5×10^{22} cm^2 / 10^{57} = 1.5×10^{-34} cm^2 per electron. In contrast, an individual electron's cross section is 6.65×10^{-25} cm^2 – over ten billion times greater. However, on average only one such star exists in its 'occupancy domain'. The total volume of such a domain, for which there is on average only one star is on the order of 2×10^{64} cm^3. So the disparity in *overall* percentage of total cross section covered by electrons when embedded in stars is even more extreme. The average distance of a line of sight to encounter one electron if all matter were in stars would be on the order of 10^{23} light years as Lord Kelvin pointed out in his resolution of Olbers' paradox.

But, if the universe were made up of uniformly dense hydrogenous plasma containing the same number of electrons on the other hand, the line-of-sight distance to encounter an electron would be much less:

$$2 \times 10^{64} \text{ cm}^3 \div 6.65 \times 10^{-25} \text{ cm}^2 = 3 \times 10^{30} \text{ cm} = 3.2 \times 10^{12} \text{ light years}$$

Of course there are variations in electron density and scattering propensity encountered in observations at cosmological distances. Each will be associated with unique total electron cross sections. Such analyses provide the key for resolving where the responsibility for thermalization lies via the scattering that ultimately reduces background radiation to microwaves. Clearly, without an intergalactic plasma medium there could be virtually no blackbody spectrum whatsoever.

quantifying 'sky cover' for uniform density cases

Let us pursue this line of reasoning more quantitatively then by assuming a uniform random distribution of density ρ per cubic centimeter of equal-sized spherical objects of cross sectional area πr_o^2, where r_o is the radius of each object. Consider a solid angle that subtends shells of uniform thickness Δr. The proportion of the total surface area of just those objects subtended in each shell would then be:

$$\alpha = \pi \rho r_o^2 \Delta r$$

There is no explicit dependence on either radial distance or solid angle here because proportionate squared factors involving them cancel. However, some of the coverage in the nth shell (and indeed increasing percentages of it) will have been occluded by objects closer to the observer and thus could not be observed.

So if we are interested in that portion of the observed field of view covered exclusively by objects in the shell at the distance r, we must subtract the amount we attribute to objects that are closer than r as suggested above. Thus, it is easier to solve the problem by working outward from the observer to more and more distant shells that we label 1, 2, 3, … n. Then the proportion in shell n included in any solid angle that is not already occluded by objects included in preceding shells we define as a(n), where:

$$a(0) = (1 - \alpha)^0 = 1$$
$$a(1) = (1 - \alpha)^1 = (1 - \pi \rho r_0^2 \Delta r)$$
$$a(2) = (1 - \alpha)^2 = (1 - \pi \rho r_0^2 \Delta r)^2$$
$$\cdots$$
$$a(n) = (1 - \alpha)^n = (1 - \pi \rho r_0^2 \Delta r)^n$$

From this derivation of a(n), we see that an exponential approximation applies for large enough values of n, i. e., when $n \gg \pi \rho r_0^2 r$, where we obtain,

$$\Delta r = r / n \rightarrow 0.$$

Then, recognizing the usefulness of changing from a(n) to a(r), we obtain in the limit as $n \rightarrow \infty$,

$$a(r) = e^{-\pi \rho r_0^2 r}$$

This employs the definition of the exponential function, which is:

$$e^{-x} \equiv \lim_{n \rightarrow \infty} (1 - x / n)^n ,$$

In order to make sure that there are no unaccounted objects occluded within a shell itself, the width Δr must be chosen such that the probability of occlusion within a shell is nil, which in the limit, will indeed be the case, of course. Curve 1 in figure 21.3 plots a(r). Curve 2, is its complement, total sky cover out to the distance r. In the figure as drawn, these curves are independent of the value of $\pi \rho_0 r_0^2$.

sky cover as closure criteria for a 'cavity surface'

Sky cover provides a basis for determining the percentage of 'closure' of a cavity surface. The rate of closure as a function of distance becomes a weighting factor pertinent to thermalization analyses. It will prove of particular significance in the redshifting environment of the intergalactic medium.

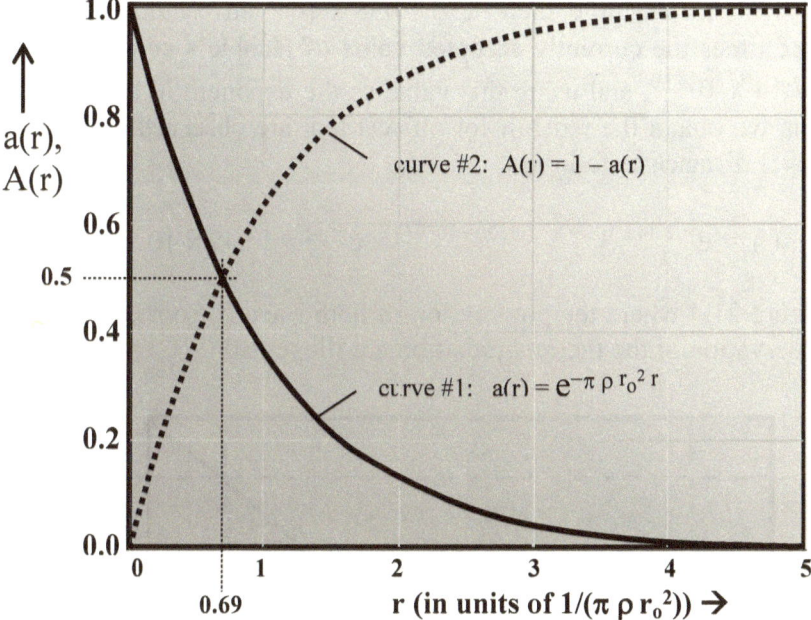

Figure 21.3: Sky coverage (curve #2) of uniformly sized objects

First we define η, which is the exponential factor in the sky cover expression as derived above:

$$\eta \equiv \pi \, \rho_0 \, r_0^2.$$

Plasma electrons have considerably larger cross sections than do the various nuclides that also populate intergalactic space. An accurate value of the electron cross section is 6.65×10^{-25} cm^2, resulting in the following:

$$\eta = 6.65 \times 10^{-25} \, \rho_e.$$

Now, let us assess the distance $R_{1/2}$ for which the percentage of sky cover would be 50 percent, i. e., $A(r) = 0.5$. So that:

$$e^{-\eta R_{1/2}} = \tfrac{1}{2}$$

Since the natural log of one half is equal to -0.69, we obtain:
$$\ln(e^{-\eta R_{1/2}}) = -0.69$$

$R_{1/2} = 0.69 / \eta \cong 1.038 \times 10^{24} / \rho_e.$

If $\rho_e = 10^{-5}$, for example, then $R_{1/2} \cong 1.038 \times 10^{29}$ cm. Multiplying this distance times the currently accepted value of Hubble's constant, i. e., $H_o = 7.14 \times 10^{-29}$, and using this value as the exponent in the redshift formula we obtain the redshift for objects that are observed at the half sky cover distance as follows:

$$(Z_{1/2} + 1) = e^{(7.14 \times 10^{-29}) \times (1.038 \times 10^{+29})} = e^{7.408} = 1.649 \times 10^3$$

See figure 21.4 where the progression of both r and Z from emission to the observation of the thermal radiation are illustrated.

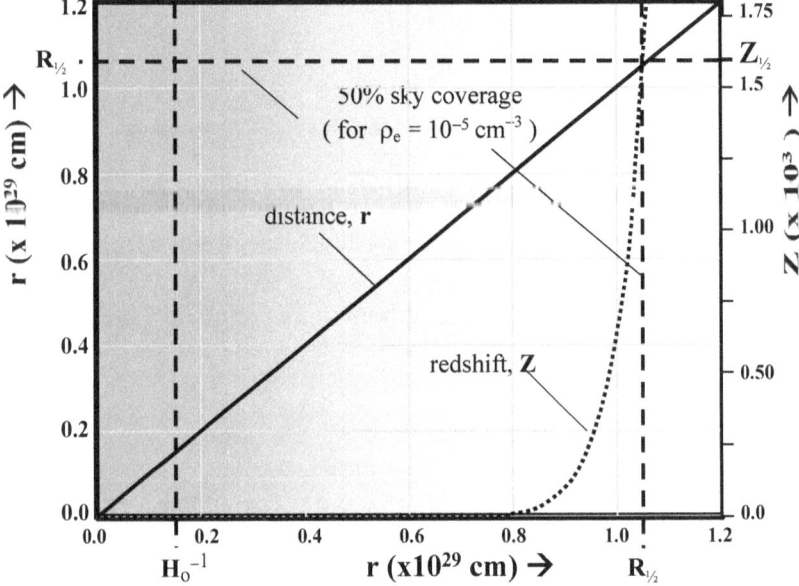

Figure 21.4: Distance and redshift to half sky cover of plasma electrons

The remoteness of the mean distance to an electron participating in the analogy to a 'cavity surface' for the vast majority of observed thermal emissions implies, of course, that much of the thermal radiation from an indefinitely extended medium would be redshifted on the order of 10^3 to 10^4 for the case we have just calculated above. Thus, because of its extremely diffuse nature in a redshifting environment, we should expect the apparent temperature of radiation to be considerably reduced from that of its emitted spectrum. This is true for any redshifting mechanism, not excluding the standard cosmological model.

To the extent that sky cover and redshift can be modeled as uniformly continuous phenomena, the proportion of solid angle subtended by electrons out to a distance r would be given by:

$$A(r) = \int_{A(o)}^{A(r)} dA(\eta, r) = \eta \int_{o}^{r} e^{-\eta r} dr' = 1 - e^{-\eta r}$$

From which the following differential equation results:

$$\frac{dA(\eta, r)}{dr} - \eta A(\eta, r) + \eta = 0$$

The function A(r) was plotted as curve #2 in figure 21.3 where $A(\infty) = 1$ as is to be expected. For a 50% surface coverage calculation if we tentatively assign a value of $\rho_e = 10^{-5}$ cm^{-3}, we would obtain a value of $\eta \cong 6.65 \times 10^{-30}$ cm^{-1}, so that $R_{1/2} \cong 1.038 \times 10^{+29}$ cm.

Clearly, the 50% sky cover situation is extremely sensitive to the value assigned to ρ_e. Using an estimate of $\rho_e = 10^{-6}$ cm^{-3}, we obtain:

$$\eta \cong 2.88 \times 10^{-7} \times 6.65 \times 10^{-25} = 1.915 \times 10^{-31} \text{ cm}^{-1} \text{ and}$$

$$R_{1/2} \cong 3.6 \times 10^{30} \text{ cm}$$

The redshift associated with this value is, of course, very appreciable:

$$Z_{1/2} + 1 = e^{7.14 \times 10^{-29} \cdot 1.038 \times 10^{+30}} \cong 10^{32}$$

For all practical purposes this is an infinite value. Thus a difference in electron density of a single order of magnitude precipitates this prodigious increase in redshift of the average line of sight distance to an electron 'surface' of more than ten orders of magnitude. However, the actual distance is less than 74 times the Hubble distance, still very appreciable, of course, but a more reasonable distance over which thermalization would occur in an extremely thin but extensive plasma.

The temperature of virtually everything in the universe – the intergalactic medium in particular – is many orders of magnitude greater than that associated with the spectrum of the microwave background radiation. In addition, the density of the universe – and again, that of the intergalactic medium in particular – is orders of magnitude less than what is directly implied by the background radiation temperature. It is certainly meaningful to inquire why such a seemingly imponderable incompatibility pertains? The standard cosmological model offers one explanation, an intermediate 'surface of last scattering', that assumes an abrupt end to scattering phenomena.

applying the analysis to thermodynamic properties

We must explore the reasons why the observed temperature of a blackbody radiation distribution emanating from an extended medium might be significantly less than would be calculated using only the kinetic temperature of the involved particles. Naturally, 'apparent' temperatures of encompassing 'surfaces' deeper and deeper in the medium from which 'cavity' radiation arises will be increasingly affected by redshift. The temperature of radiation from a single expanding (or otherwise redshifted) surface as hypothesized in the standard cosmological model would be reduced as the inverse first power of the redshift:

$$T_{rad}(r) = T_{sur}(r) / (Z(r) + 1) = T_e \, e^{-H_0 \, r},$$

A less complex alternative is that the ubiquitous microwave background radiation is merely the expected radiation associated with high temperature plasma that has been redshifted by varying degrees. The analyses concerning sky cover shed some light on related questions as preparation for a fuller explanation.

So we come now to the issue of using the weighting factor for integration in order to correctly assess effects that are dependent upon a redshift that occurs over the distances to component portions of the surrounding 'surface' that ultimately effect the total 'cavity surface' applicable to thermodynamic analyses as discussed in chapter 7. We illustrate what is involved in figure 21.5. Where the local thermal medium emits thermal (but not necessarily blackbody) radiation at high temperature that is ultimately integrated into a blackbody spectrum of a much lower temperature. This integration procedure is described by Bonn (2009). It is reasonable to assume that over any *cosmological* distance the intergalactic medium would be statistically uniform such that a 'cavity surface temperature' would equal the kinetic temperature T_e, i.e., that $T_{sur}(r) = T_e$, a constant. Then for redshift derived from scattering as we propose in particular, the following applies:

$$T_{rad}(r) = T_e / (Z(r) + 1) = T_e \, e^{-H_0 \, r},$$

Just as an example, if we were to use an estimate of the electron density of $\rho_e = 10^{-5}$ cm^{-3} from which to calculate a half-sky cover for

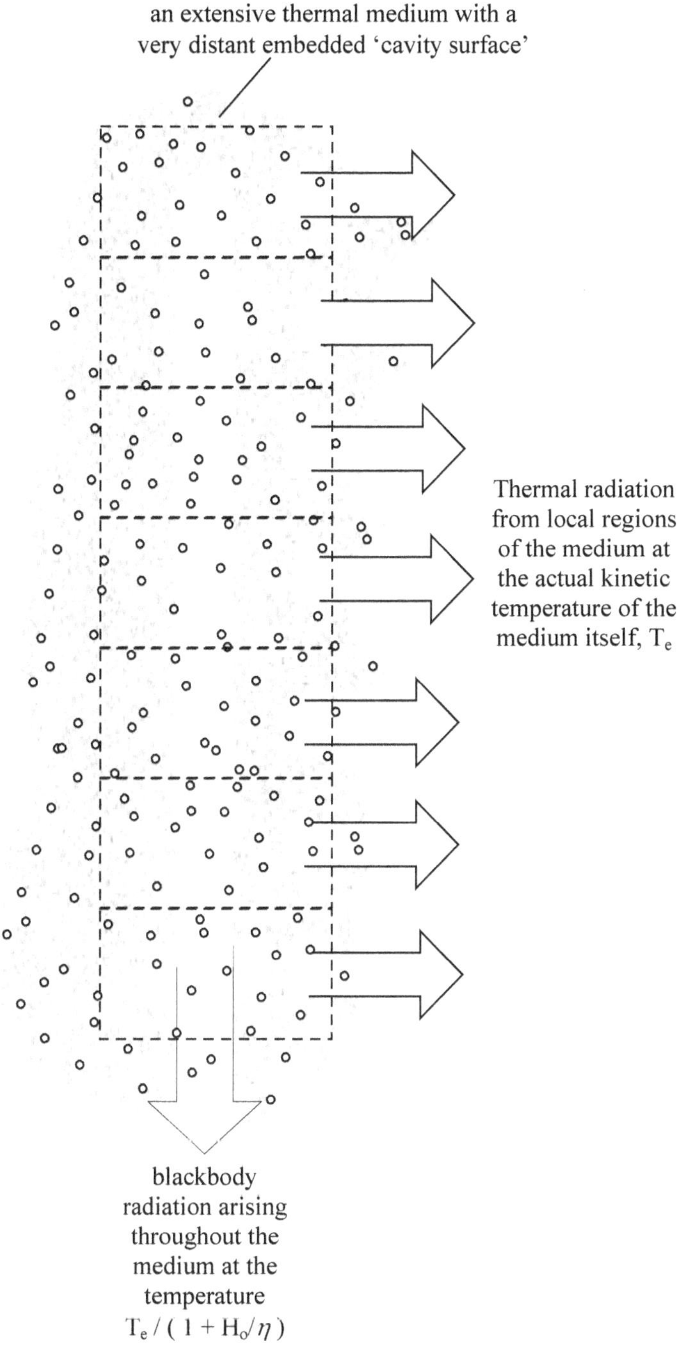

an extensive thermal medium with a
very distant embedded 'cavity surface'

Thermal radiation
from local regions
of the medium at
the actual kinetic
temperature of the
medium itself, T_e

blackbody
radiation arising
throughout the
medium at the
temperature
$T_e / (1 + H_o/\eta)$

Figure 21.5: **Blackbody radiation resulting from integration of thermal radiation from all depths of the medium**

the thermal cavity and a kinetic temperature of the medium of 4.50 x 10^3 K, the apparent temperature of radiation originating at particles at the 'mean' distance of a surrounding 'cavity surface' would be given by:

$$T_{rad}(r_{1/2}) = T_e / (Z_{1/2} + 1) \cong 4.50 \times 10^3 \text{ K} / 1.649 \times 10^3 \cong 2.725 \text{ K}$$

These assigned parameter values would, in fact, determine a background radiation temperature of 2.725 K. However this is only because we specifically selected values of ρ_e and T_e to produce that result. Although such fudging of data may serve for some as explanation in the standard model where such an approach is used to define a "surface of last scattering", it does *not* legitimately accrue credibility on that account for several reasons. Not least of these reasons is that validity of the analysis depends upon the cavity 'surface' being *complete*, i. e., that it possess no closer obstacles or holes through which radiation of a different temperature (or none at all) could have arisen. This criterion is not met by the scenario to which it is applied in the standard (or any other) model.

If a value of electron density had been chosen such as, $\rho_e = 10^{-7}$ cm^{-3}, an associated 'mean' radiation temperature would have been only insignificantly above absolute zero unless we were to assign a truly astronomical kinetic temperature at the surface. But a temperature of any such incomplete particular 'surface' is of very little significance in any case. What *is* important, however, is the composite effect of all the conceptual partial 'cavity surfaces' throughout the universe that produce closure. A prorated average of the redshifted radiation temperatures, $<T_{rad}(R)>$ for $0 \leq R \leq \infty$ from all partial radiating 'surfaces', is what is significant in attempts to predict background radiation in an extensive substance sufficiently diffuse for redshifting to be significant.

This can be assessed by integrating the effects of these radiation temperatures from individual 'surface areas' throughout the extensive regions, each providing a portion of the total surface area of a legitimately realistic cavity 'enclosure'. In a medium of infinite extent, the total effect from beyond a distance, R can be obtained by integration using the following formula:

$$< T_{rad}(r > R) > = \int_R^\infty T_{rad}(r) \, dA(\eta, r) = \int_R^\infty \frac{T_e \, \eta \, e^{-\eta r}}{Z(r) + 1} dr = \frac{T_e \, e^{-(\eta + H_o) R}}{1 + H_o / \eta}$$

Here we have addressed the closure problem with the weighting factor $A(\eta, r)$ that was derived above.

The generality of expressions for $<T_{rad}(r>R)>$ decreases from left to right. The first applies to any explanation of background radiation in a redshifting environment independent of the redshift mechanism. The next applies to the extent that the actual density and temperature are statistically constant. However, since these parameters increase with lookback distance in the standard cosmological models, a somewhat different final formulation would be required. Such analyses are definitely required by that model and it is considerably remiss of cosmologists not to have incorporated a critical assessment of the impact of this phenomenon. The expression at far right incorporates the logarithmic/exponential redshift-distance relationship derived specifically based upon the scattering in a thermodynamic system.

If we use this formula to determine the *overall average* radiation temperature (i. e., for $r > 0$) when looking out through the intergalactic medium from any observation point, we obtain:

$$<T_{rad}(r>0)> = \frac{T_e\, \eta}{(H_o + \eta)} = \frac{T_e}{(1 + H_o/\eta)}$$

So we have come finally to a determination of the relationship of the kinetic and radiational temperatures in a redshifting medium. Of course we know what this observed radiation temperature is for our universe. It is 2.725 K.

What should have been obvious, but because of bad habits seems to come as an extremely profound realization, is that background radiation does not impose its own temperature value on the rest of the universe. In fact, the ratio of the kinematic and blackbody radiation temperatures in an extensive medium in equilibrium is given as:

$$T_{k/r} \equiv T_{kin}/T_{rad} = T_e/<T_{rad}(r>0)> = 1 + H_o/\eta$$

Here we have substituted values for the kinetic and the radiational temperatures. By plugging in the values for Hubble's constant and the electron cross section in terms of electron density, to obtain:

$$T_{k/r} = 1 + 1.074 \times 10^{-4}/\rho_e$$

And importantly, we have that:

$$T_e = <T_{rad}(r>0)> \times T_{k/r} = 2.725 \times (1 + 1.074 \times 10^{-4}/\rho_e)$$

Clearly, whenever the density of the medium is appreciable, the kinetic and background blackbody temperature will be equal. It is this constraint that has previously been missing from claims of their being identical. This formulation is illustrated in figure 21.5.

For even an extensive medium that is more dense than $\rho_e = 10^{-3}$ cm^{-3} (which is much more vacuous than any medium ever encountered in any laboratory on earth) the kinetic temperature of the particles in the medium and the temperature of radiation given off will be identical. But although this has typically been expected to be true in *every* case, it is most definitely *not* the case for the intergalactic medium as clearly illustrated in figure 21.6.

parameter averaging appropriate to H$_o$ and η

Viable points on the plot in figure 21.6 might seem to imply a basic incompatibility with the scattering redshifting mechanism requirements defined earlier. Applied directly as operative averages of electron density and kinetic electron temperature of the intergalactic medium, these values do not accord well with the constraints we established earlier for a plasma scattering mechanism assuming that that is indeed responsible for cosmological redshift as proposed here.

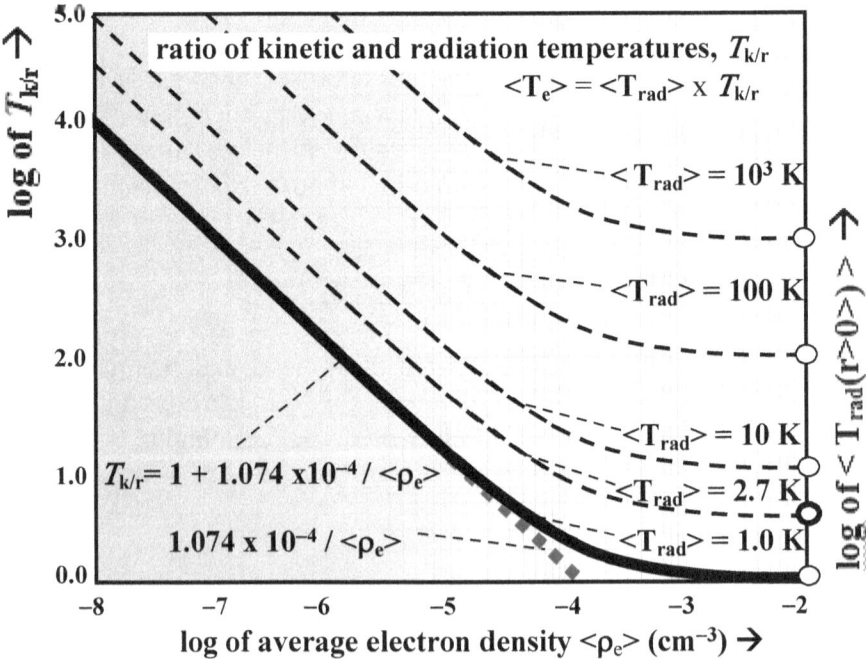

Figure 21.6: Log of ratio of average kinetic and radiation temperatures vs. average electron density of an extended thermal medium, showing implication for various radiation temperatures

We concluded the previous chapter with the explanation that for cosmological redshift to result as a direct consequence of forward scattering required that $H_o \cong 1.719 \times 10^{-32} <T_e \rho_e>$. Therefore to effect an accurate fit with the Hubble constant, we had to have that the average product of the plasma electron kinetic temperature and density had to be $<T_e \rho_e> \cong 4.13 \times 10^3$ K cm^{-3}. This would certainly seem to be in conflict with what we have just found above to match the temperature of microwave background radiation. In this case also we must deal with averages, but in this case averaged *separately*, so we have:

$$< T_e > < \rho_e > \ = \ 2.725 \times (< \rho_e > + 1.074 \times 10^{-4})$$

And since we know that $< \rho_e >$ is much less than 1.074×10^{-4} cm^{-3}, we have that the product of averages is as follows:

$$< T_e > < \rho_e > \ \approx 2.93 \times 10^{-4}$$

However, we now know also that the redshift mechanism implies that the average of the product $< T_e \rho_e > \cong 4.13 \times 10^3$ K cm^{-3}. Together these two relationships obviously imply that:

$$\frac{< T_e \rho_e >}{< T_e > < \rho_e >} \ \cong 1.42 \times 10^7$$

This equation specifies a ratio of two types of averaging, valid to the degree to which the assumptions of our analysis apply.

Where extremes exist in the parameter values, the average of the product of those parameter values differs considerably from the product of the averages of those same parameters.

In order for there to be an appreciable redshift, the mechanism of scattering redshift depends intimately upon the requirement for a high degree of ionization and on a sufficiently high density to result in extinction occurring at reasonably short intervals. But this is just a statistically averaged condition as we discussed in concluding the previous chapter. This averaging accommodates extensive variation, compatible with redshift survey variations, while still matching the observed redshift phenomena over cosmological distances. The vast majority of the redshifting produced by this mechanism is on radiation passing through intermediate galaxy clusters of which there are billions. In rich cluster cores both plasma density and temperature are much greater than their corresponding averages for the intergalactic medium as a whole.

With regard to sky cover (or 'cavity surface') we are naturally interested in density averaged separately from temperature. As we discussed above, variations in matter density produce very different cross sections for the electrons involved. In particular they produce extremely different distances to closure of a 'cavity' surface. We saw that stars and other compressed matter generally, although possibly containing most of the baryonic matter in the universe, do not contribute in any substantial way to the thermodynamics of the universe as a whole. Thus also, the relatively dense but localized intracluster plasma will not contribute as much proportionately as the more dispersed plasma in the outskirts of, and between, such rich clusters. By far the majority of thermalization will occur in intergalactic regions at great distances from clusters. Distances to the large majority of the overall cavity surface will be determined by regions with the more dispersed electron densities.

Compatibility of H_0 and η, both of which are determined in large part by plasma density, is achieved by recognizing that they are based on unique averaging processes primarily involving different regions of intergalactic space. The former averages a product that places emphasis on regions of much larger values of both temperature and density, the latter averages the distances to a cavity surface determined exclusively by the cross section contribution of electrons, and therefore, placing emphasis on the inverse of electron density. These differences, therefore, accrue values based primarily on contributions from very different regions of space.

By such analyses it is clear that the temperature of background radiation does not impose the same characteristic temperature (or even anything similar to that) of the blackbody profile onto the kinetic temperature of material aspects of the extended redshifting medium from which that radiation derives.

...it is clear that the temperature of background radiation does not imply that that characteristic temperature must be the same as the kinetic temperature of material aspects of the extended redshifting medium from which that radiation derives.

22: The Origins and Profound Ramifications of Irreversibility

"...we described some difficulties in the microscopic theory of irreversible processes. Its relation with dynamics, either classical or quantum, cannot be simple, in the sense that irreversibility and its concomitant increase of entropy cannot be a general consequence of dynamics. A microscopic theory of irreversible processes will require additional, more specific conditions. We must accept a pluralistic world in which reversible and irreversible processes coexist. Yet such a pluralistic world is not easy to accept."[28]

"There can be no doubt that irreversibility exists on the macroscopic level and has an important constructive role, as we have shown... Therefore there must be something in the microscopic world of which macroscopic irreversibility is the manifestation."[29]

We have considered the previous attempts to explain the origin of irreversibility. They have all failed. But we persisted because we knew there had to be a "manifestation" at the submicroscopic level. There is.

Boltzmann's investigations considered all possible collisions of molecules in a gas. From this he was sure that his 'H theorem' proved that a distribution of molecular velocities would ineluctably be driven to the Maxwell-Boltzmann distribution. But in fact he had only shown that this distribution is indeed associated with equilibrium conditions.

---

[28] Prigognine and Stengers, p.257
[29] Ibid, 258

The number of molecules with component velocities in each of three dimensional directions remains unchanged after each collision except that they are reversed – the epitome of a reversible interaction. There is no number of completely elastic collisions that changes the distribution of relative velocities of molecules in a gas by one iota. Nor did he shed light on how thermal radiation arises in a thermodynamic system.

There is no number of completely elastic collisions that changes the distribution of relative velocities of molecules in a gas by one iota.

Einstein's derivation of the Planck distribution of blackbody radiation associated with thermodynamic systems in equilibrium failed in similar ways to shed any light on irreversibility. Although demonstrating that the resulting distribution would remain compatible with the Maxwell-Boltzmann distribution by means of momentum transfers to the kinetic motions of the molecules in the system, he did not use his own relativistic Doppler formula in this endeavor, and because of that simplification he failed to identify a source of irreversibility.

Einstein's failure to use his own relativistic Doppler formula allowed irreversibility to slip through his fingers.

Statistical mechanics addresses the combinatorics of permutations that are inevitable concomitants of random distributions. Clearly there is an irreversible tendency to distribute an increasing amount of the total energy among more and more component particles. Clearly there are many more ways for particles to possess small portions of the total amount of energy than there are for them to possess larger portions of the total amount. But this does not in any way suggest mechanisms by which these tendencies might arise. The discipline is based exclusively on likelihood and is left vulnerable to suppositions of ridiculous *possibilities* that, for example, all of the molecules of air in a room could end up in one corner of that room defying the conservation laws. There is physically no way for that to occur.

Clearly there are many more ways for particles to possess small portions of the total amount of energy than there are for them to possess larger portions of that total amount.

root causes of irreversibility at the submicroscopic level

Virtual 'reversible' behavior is associated with slow motion. There is very nearly complete reversibility when pistons move slowly enough. With rapid motion second order effects begin to become apparent and with them, irreversibility becomes a measurable reality.

There is a time-honored adage that all of the laws of physics at the submicroscopic level are reversible. They aren't. With the advent of Einstein's relativity and

There is a time-honored adage that all of the laws of physics at the submicroscopic level are reversible. They aren't!

quantum mechanics, any excuse for believing that went away. The development of those theories at the beginning of the 20th century began what is denominated 'modern physics'. And with it the laws that describe submicroscopic behavior changed forever. Whenever particles and radiation interact, that interaction is irreversible. The impact is miniscule on a single interaction, but it is unilateral and cumulative. The interactions between particles and radiation provide the mechanism whereby the redistribution of energy is increasingly spread among more and more of the constituent particles of a thermodynamic system.

We have seen that there is incompatibility of the conservation laws for kinematic interactions of particles and those same laws associated with radiational interactions. The expression for the conservation of energy of a particle is nonlinear with respect to the expression for the conservation of momentum of those particles. There is on the other hand a linear relationship between

The expression for the conservation of energy of a particle is nonlinear with respect to the expression for the conservation of momentum of that particle. There is on the other hand a linear relationship between those conservation expressions for a photon of radiation. That difference is the root cause of irreversibility.

conservation expressions for a photon of radiation and that difference is the root cause of irreversibility. An interaction between particulate matter and radiation inevitably results in irreversible behavior on that account.

entropy at the submicroscopic level

The descriptive facts of entropy link it directly to the concept of irreversibility. Now since we have identified irreversible interactions that occur at the submicroscopic level of our reality, we must show that this behavior is in fact the reduction of entropy to the submicroscopic level in order to satisfy our scientific reductionist agenda.

First of all entropy demands that energy flow from the higher to the lower energy segment of a system or combined systems. It is only through the mediated or scattering interactions of radiation with the particulate aspect of systems that that happens. Elastic collisions of particles can never transfer energy out of the particulate segment of a system, nor can they even change the distribution of permutations of the energy among the component particles. One particle takes on the energy of another with no significant change whatsoever in the total distribution of energy in the system nor of its permutations.

However, with the mediated interactions resulting from radiation exchanges, the excessive energy of one molecule relative to that of another is reduced in the exchange. This results in the two molecules having more nearly similar amounts of energy and both of them having less than that of the emitting molecule. This sharing of energy does, in fact, produce the increase in permutations involving smaller amounts of the total energy just as statistical mechanics has suggested must be the case with an associated increase in entropy. Thus we have identified the specific mechanism for bringing this about.

> *Sharing of energy increases the number of permutations that involve smaller amounts of the total energy which we have learned must be associated with an increase in entropy. Thus we have identified the specific mechanism that causes entropy to increase.*

The mathematical combinatorics of statistical mechanics merely describes the results of irreversible redistribution of permutations of the allowed energy allocations among particles. But again there is in that discipline no attempt to explore the mechanisms by which the redistribution occurs.

Einstein's treatment of the quantum theory of radiation elaborates how energy stored in molecules is transferred to the radiation segment so that rather than being locked up within the particulate segment, it is redistributed in transit to the radiation distribution. But there is no demonstration of irreversibility here. The photons of radiation emitted by either of his defined methods are assumed to be capable of adhering to particulate matter again with only a reversible classical momentum exchange between segments. The treatment is probabilistic so that the mechanisms whereby exchanges occur are obscured from scrutiny.

This is where clarification of Einstein's disagreement with Walter Ritz concerning the origin of irreversibility could have had a significant impact on his treatment. Ritz saw the distinction between retarded and advanced wave solutions to Maxwell's equations as a cause of irreversibility. He was close. So in the end, although having made a major step forward, Einstein left the issue of irreversibility unresolved. It is resolved by the details of the mechanism whereby only certain exchanges are allowed, and these happen to be irreversible.

heat at the submicroscopic level

Exclusive attribution of 'heat' to molecular motions is misguided in as much as being hit by a fast particle would not feel 'hot'. It would probably hurt like being hit with a small rock at high speed. Being irradiated by high frequency radiation burns – like being burnt by fire. So why is radiation not a major part of the answer to "What is heat?" It is in fact the *major* aspect of any correct answer in this regard.

We acknowledge that entropy is associated with the transfer of heat' from a system with the most heat to the one with the least heat; it is radiation that does that, not particle collisions. What occurs at the submicroscopic level traces directly to 'heat' as it should.

> *Being hit by a particle probably hurts. Being irradiated by high frequency radiation burns – like being burnt by fire. So why is radiation not a major part of everyone's answer to "What is heat?"*

closing loops in the models of submicroscopic behavior

We mentioned that the major attempts to model submicroscopic behavior failed to close all the loops so that the veracity of every attempt remained in question. In closing those loops there is the transfer of energy across boundaries of molecules (particulate matter in general) transferring energy and momentum to radiation. This occurs in the mediated exchanges of photons of radiation between molecules. This, certainly not elastic collisions, is what brings about the Maxwell/Boltzmann distribution of particulate kinetic energies. Elastic collisions only maintain it.

The transfer of energy and momentum to particles happens as a result of scattering of radiation by charged particles as well. In this way the energy of high frequency radiation is transferred to particles as a part of the process of redshifting in thermal media. This is the process of thermalization that brings the thermal radiation to the stable blackbody form.

Figure 22.1 illustrates the mechanisms that produce the necessary and sufficient conditions for the respective distributions of energy. If they are not matched, then an adjustment occurs naturally as a part of the loop closure that initially introduces and thereafter maintains a system in equilibrium.

closing loops at the top level of our universe

But we have not addressed the whole story. There is a universe 'out there' that is in a dynamic stasis maintained by the opposing forces of both gravity and thermodynamics. Rejecting the 'astrophysical trend' as an origin of irreversibility does not eliminate high level macroscopic processes from consideration in discussions of thermodynamics.

Rejecting the 'astrophysical trend' as an origin of irreversibility does not eliminate high level macroscopic processes from consideration in discussions of thermodynamics.

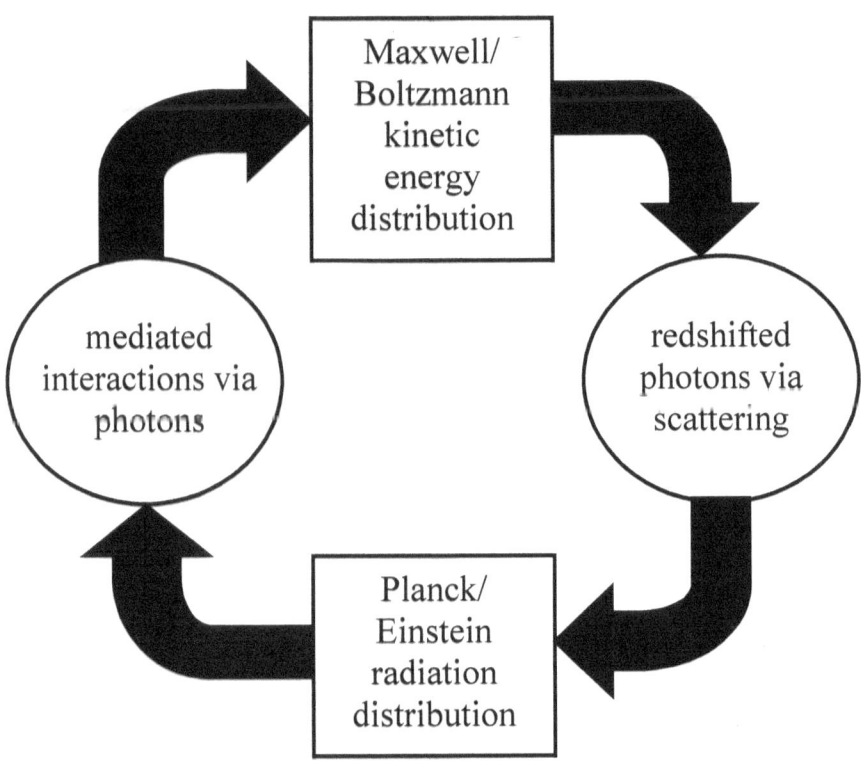

Figure 22.1: **Energy distributions being equalized by irreversible processes that drive a system to equilibrium**

Entropy is sometimes interpreted as the level of disorder of a system, or that part of a system that cannot be used to produce physical work. In another sense, entropy is concerned with the amount of information required to describe a system. A disorganized system requires more information to describe it accurately than does an orderly one. The affirmation that entropy can never decrease implies that the requisite physical description of a system cannot be reduced, i. e., information cannot be destroyed. Expetives may *not* be deleted.

And yet, black holes for which there is undeniable evidence seem to do just that. An adage that is on a par with the no free lunch maxim associated with the second law of thermodynamics is that 'black holes have no hair'. This is because a black hole involves no subtleties of description; there is their mass, their electronic charge, and their rotation – nothing more. What this means is that if a thermodynamic system with considerable entropy were to be swallowed by a black hole, the entropy of that system would virtually disappear and entropy of the global system that includes the black hole would decrease.

> *If a thermodynamic system were swallowed by a black hole, the entropy of that system would virtually disappear and entropy of the entire universe, including the black hole, would decrease.*

The thermodynamics of black holes is a subject of some interest to the physics community on this account. We will not proceed further into this area of discussion, but it is an interesting area the reader may wish to investigate further. A way out of this dilemma as it has been presented has seemed to be Hawking radiation by which a black hole could eventually effervesce itself away over a period of time of on the order of 10^{85} years. But... during the hiatus? Entropy was decreased.

Be that as it may, we now know of gamma ray bursts with energies so great that they recreate conditions that emulate what has been attributed to a 'big bang' origin of the entire universe. More recently even more direct evidence of black hole eruptions has been obtained.[30] It seems reasonable to assume that these eruptions are indeed that of black holes, spewing forth the primordial hydrogenous plasma of the intergalactic medium. At the temperatures prevalent during gamma ray bursts thermonuclear processes occur whereby the 24% helium by mass is maintained as a ubiquitous constant in mass distributions throughout our universe.

The conversion of hydrogen into helium results in the generation of high energy radiation as a part of balancing conservation equations. In fact, as described by Bonn, this amount of hydrogen conversion would produce a radiation energy density throughout the universe of ρ_{rad} as follows:

$$\rho_{rad} = 4.169 \times 10^{-13} \text{ ergs per cm}^3$$

This is precisely the amount of energy per cm^3 that is currently observed in the microwave background radiation. This cannot be a

---

[30] http://phys.org/news/2015-06-chandra-evidence-serial-black-hole.html#nRlv

coincidence. These are simply the manifestly observed facts of our universe.[31] In figure 22.2 we reproduce a diagram from Bonn (2011) illustrating the loop closures of phenomena at the highest levels of a complete description of our universe. This universe is obviously not static, nor can it be represented by an the invalidated 'steady state universe' model. But it is in a 'stationary state', meaning that at the highest level nothing changes. People come and go, planets come and go, even stars come and go, and galaxies, and even black holes but the universe is here for the long haul. There is no 'astrophysical trend' as a demise of the universe as a whole at the highest level.

> *The universe is obviously not static, nor can it be represented by an erroneous 'steady state universe' model. But it is in a 'stationary state', meaning that at the highest level nothing changes.*

the big picture with regard to irreversibility and entropy

Irreversibility and entropy are major concepts concerning the phenomena we observe in the universe we inhabit. They are inextricably linked. One implies the other, but they are not equivalent. Irreversibility is a low level concept that most accurately applies to individual interactions; interactions can either be reversed or they cannot. Entropy on the other hand applies most specifically to thermodynamic systems as a whole. We have discussed the individual interactions whereby irreversible behavior makes its entry. But what is the associated story for entropy from a perspective of what occurs at the highest level of our universe?

The entropy of any system must somehow drop to near zero if it is swallowed by a black hole. But then what happens with the subsequent eruption of that black hole, whether a primordial universe-in-a-pinhead gigantic blackhole with a one-time-big-bang or with a gamma ray burst that signals a somewhat smaller event, entropy proceeds to (once again) continuously increase. Facts now seem to reveal exactly that latter scenario, where from a universal perspective overall entropy actually decreases significantly in a first phase but, may very likely once again reinstigate continuous increases per expectation. All of this is a direct affront to what has been accepted as the undeniable facts associated with the second law of thermodynamics with entropy never decreasing. Of course the entropy of any system we could actually observe in our mundane world would always continuously increase.

---

[31] Bonn (2009) pp. 37-40.

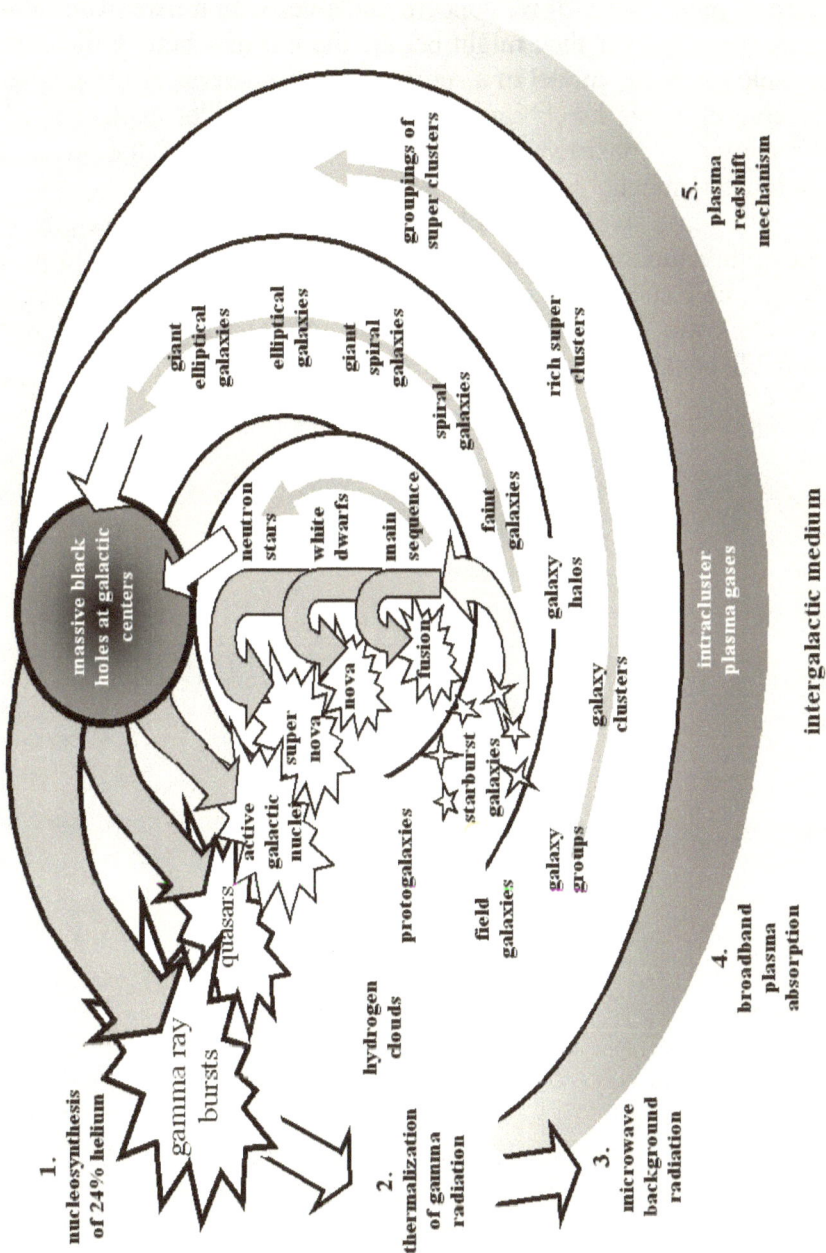

Figure 22.2: The universe whose stasis is maintained by loop closures

The submicroscopic processes associated with gravitation whereby entropy might actually decrease to near zero when a system collapses into a black hole are not addressed here, because that was not a primary concern of ours. Nor will we concern ourselves with it here. But those phenomena, whatever they might be, are the means whereby the outer loop is closed in any model of a stationary state universe. Entropy, like everything else, has its ebb as well as flow. A similar explanation is required of any consistent explanation by advocates of the standard cosmological model.

In any case we have solved the mystery of the origin of irreversibility and the ineluctable increases in entropy. That origin is in the irreversible submicroscopic processes that do in actual fact occur as we have shown. The ramifications of the tiny effects associated with irreversible behavior are indeed profound.

The Origin of Irreversibility

Appendix

Frequently encountered constants

a. physical constants applicable to the universe

c (speed of light in a vacuum) = 2.9979 x 10^{10} cm/sec.

h (Planck's constant) = 6.626 x 10^{-27} erg sec

k (Boltzmann's constant) = 1.380 x 10^{-16} ergs/degree kelvin

σ (Stefan-Boltzmann constant) = 2.268 x 10^{-4} erg-cm^{-2}-deg^{-4}

N_A (Avogadro's number) = 6.0225 × 10^{23}

H_o (Hubble's, constant) $\cong$ 7.14 x 10^{-29} cm^{-1}

$\cong$ 67. 4 km sec^{-1} Mpc^{-1} / c

ρ_o (Einstein's critical density) $\cong$ 8.0 x 10^{-30} gm cm^{-3}

ρ_{mu} (mass density of the universe) $\cong$ 5.49 x 10^{-31} gm cm^{-3}

$\rho_{\mu bb}$ (microwave background density) = 4.176 x 10^{-13} ergs cm^{-3}

$T_{\mu bb}$ (microwave background radiation temperature) = 2.725 K

b. constant properties of objects within the universe

e (electronic charge) = 4.80 x 10^{-10} stat coulombs

m_e (electron rest mass) = 9.109 x 10^{-28} gm

r_e (classical electron radius) = $e^2/ m_e c^2 \cong$ 2.82 x 10^{-13} cm

m_p (proton mass) = 1.672621637 x 10^{-24} gm

m_n (neutron mass) = 1.67492729 x 10^{-24} gm

$M_\odot$ (mass of the sun) $\cong$ 2.0 x 10^{33} gm

$R_\odot$ (radius of the sun) $\cong$ 6.9550x 10^{10} cm

c. unit conversion constants

1 light year (distance) = 9.4606 x 10^{17} cm

1 Angstrom = 10^{-8} cm

1 eV (electron volt) = 1.602 x 10^{-12} ergs $\rightarrow$ 1.161 x 10^4 K

1 MeV (energy) = 10^3 KeV = 10^6 eV

The Origin of Irreversibility

Bibliography

1. Mark Birkinshaw, "The Sunyaev-Zel'dovich Effect", <u>Physics Reports</u>, 310, 97 (1999)

2. Bonn, R. F., <u>The Aberrations of Relativity</u>, ISBN 978-0-6151-9781-4, Vaughan Publishing, Seattle (2008)

3. Bonn, R. F., <u>Cosmological Effects of Scattering in the Intergalactic Medium, 2nd Edition</u>, ISBN 978-0-578-02625-1, Vaughan Publishing, Seattle (2011)

4. M. Born and E. Wolf, <u>Principles of Optics, 6th</u> Ed., Pergamon Press, New York, NY, pp. 93-96 (1980)

5. K. Brecher, "Is the Speed of Light Independent of the Velocity of the Source?" <u>Phys. Rev. Letters,</u> Vol. 39, N0. 17, pp. 1051-1054 (24 October 1977)

6. Ari Brynjolfsson, "Redshift of photons penetrating a hot plasma", (2005), arXiv:astro-ph/0401420v3

7. John E. Carlstrom, Gilbert P. Holder, and Erik D. Reese, "Cosmology With The Sunyaev-Zel'dovich Effect," <u>Annu. Rev. Astron. Astrophys</u>, 40:643–80 (2002)

8. Cowen, "Safe from a heavenly doom: gamma-ray bursts not a threat to Earth", <u>Science News</u>, pp 88-89 (2006)

9. L. L. Cowie and S. C. Perrenod, "Origin and distribution of gas within rich clusters of galaxies: The evolution of cluster x-ray sources over cosmological time scales", <u>The Astrophysics Journal,</u> 219, 254 (1978)

10. J. G. Cramer, "Generalized Absorber Theory and the Einstein-Podolsky-Rosen Paradox," <u>Phys. Rev. D</u>, 22, 2, 362-376 (1980).

11. J. G. Cramer, "The Transactional Interpretation of Quantum Mechanics," Rev. Mod. Phy., 58,3, 647-687 (1986)

12. J. G. Cramer, "Generalized Absorber Theory and the Einstein-Podolsky-Rosen Paradox," Phys. Rev. D, 22, 2, 362-376 (1980).

13 R. W. Ditchburn, Light, 2nd Ed., Interscience, New York, 1963 (p. 681)

14 John Earman, Proceedings of the Biennial Meeting of the Philosophy of Science Association, Vol. 2, (1986)

15 Albert Einstein, "On the Quantum Theory of Radiation," Sources of Quantum Mechanics, Dover, New York, 1967. (p. 64)

16 M. J. Geller and P. J. E. Peebles, "Test of the Expanding Universe Postulate," The Astrophysical Journal, Vol. 174, pp. 1-5 (15 May 1972)

17 George Gamow, Thirty Years that Shook Physics – the Story of Quantum Theory, Dover, New York (1966)

18 John Gribbin, Q is for Quantum – An Encyclopedia of Particle Physics, The Free Press, New York, P.411 (1999)

19 Thomas S. Kuhn, The Structure of Scientific Revolutions, 3rd Ed., Univ. of Chicago Press, Chicago (1962, 1970, 1996)

20 Leonard Loeb, Kinetic Theory of Gases, McGraw-Hill, New York, (1927)

21 G. N. Lewis, "The Nature of Light", Proceedings of N. A. S., 12, 22-29 (1926)

22 Karl Popper, Conjectures and Refutations, ISBN 0-415-28594-1, Routledge, New York (2002)

23 R. L. Oldershaw, "Cosmology theory compromised," Nature, Vol. 346, p 800 (19 July 1990)

24 I. Prigogine and I. Stengers, Order out of Chaos, Bantam Books, Toronto (1984)

25 Erwin Schrödinger, <u>What is Life?</u> , Doubleday, New York, 1956.

26 William J. Sidis, "The Paradox," <u>The Animate and the Inanimate</u>, Boston: Badger, 1925. This volume is also available on the internet at <u>http://www.sidis.net/ANIM4</u>.

27 H. C. Van Ness, <u>Understanding Thermodynamics</u>, Dover, New York, 1969 (p. 28) But, of course we are more interested in the submicroscopic level we save till later.

28 R. Fred Vaughan, <u>The Relativity of Visual Observations</u>, ISBN 1507641788 and ISBN 9781507641781, Vaughan Publishing, Seattle (2015)

29 Steven Weinberg, "Can Science Explain Everything? Anything?" <u>The New York Review</u>, May 31, 2001

30 J. A. Wheeler and R. P. Feynman, "Interaction with the Absorber as the Mechanism of Radiation," <u>Rev. Mod. Phy.</u>, 17, 157 (1945)

31 J. A. Wheeler and R. P. Feynman, "Classical Electrodynamics in Terms of Direct Interparticle Action," *Rev. Mod. Phy.*, 21, 425 (1949)

32 E. Wolf and D. N. Pattanayak, "General Form and New Interpretation of the Ewald-Oseen Extinction Theorem," <u>Optics Communications,</u> Vol. 6, No. 3, pp. 217-220 (November 1972)

33 E. Wolf, "Non-cosmological redshifts of spectral lines," <u>Nature,</u> Vol. 326, No. 6111, pp. 363-365 (26 March 1987)

34 E. Wolf, J. T. Foley, and F. Gori, "Frequency shifts of spectral lines produced by scattering from spatially random medium," <u>J. Optical Soc. Amer. A,</u> Vol. 6, No. 8, pp. 1142-1149 (August 1989)

35 E. Wolf, "Invariance of the Spectrum of Light on Propagation," <u>Phys. Rev. Letters,</u> Vol. 56, No. 13, pp. 1370-1372 (31 March 1986)

36 E. Wolf, "Recent Work on the Ewald-Oseen Extinction Theorem," <u>Atomic and Molecular Optics, Proc. of the 1971 Rochester Symp.,</u> J. H. Eberly, Ed., Univ. of Rochester, NY, pp. 55-67 (March 1971)

37 E. L. (Ned) Wright, "Measuring the Curvature of the Universe by Measuring the Curvature of the Hubble Diagram", from <u>Ned Wright's home page</u>, (May, 2008)
http://www.astro.ucla.edu/~wright/sne_cosmology.html

Other books by the author

Not Julie - a Trilogy   by R. F. Vaughan
This Faustian tale pits an aging Ray Bonn against the scientific establishment, the premier slugger in major league baseball, and romantic temptations. He seems to overcome all obstacles, some of which were shoved into his path by a charming young physicist, Lesa Landau. She always acts in what she perceives as Ray's best interest but he is not always convinced of that. Together they explore the rationale for the saying, "There is no free lunch", and personally experience the ramifications of that glib saying. They also collaborate on their other scientific interests in relativity and cosmology.

They age and die tragically with their offspring challenged by their own most frustrating issues.

This volume contains over 700 pages. It can be purchased at CreateSpace.

The Relativity of Visual Observation   by R. F. Vaughan
This monograph provides a reformulation of relativity theory that emphasizes a direct relationship between the observations of observers in relative motion. It draws a clear distinction between actualized observation and inferences from theoretical constructs that cannot be observed. The focus is on visual observations rather than abstractions that have traditionally been supposed to constitute observations in relativity.

Roger Penrose demonstrated that besides the Lorentz transformation a second 'transformation of the field of vision' is required to transform visual observations between relatively moving observers. This dual transformation set transforms observed circles back into circles refuting Einstein's prediction that spheres would "appear oblate" to an observer in relative motion. This monograph proceeds further to examine the appearance of wall clocks under similar circumstances by applying the same 'transformation of the field of vision'. The approach provides a time stamp on a neighborhood of transformed events which brings time dilation into question as well.

As points of departure, alternative hypotheses are presented that provide different solutions to this 'problem' with interpretations of the inevitable spatial and temporal disparities which consistently predict experimental observations. Finally a single transformation with no intermediary metaphysical distractions is derived by embracing observable aberration effects on the electric and magnetic fields of electrodynamics. It provides a physical basis (rather than mathematical contrivance) for understanding the ostensible effects of relative motion. It accommodates the essential features of covariance, generalization, and compatibility with quantum theories.

This monograph can also be purchased at CreateSpace.

R. F. Vaughan has chosen the name of the fictional characters, Ray Bonn and Lesa Sorrensen, from his novel *Not July* as a pseudonym for the author in the current book. In these other of his scientific publications he has used the pseudonym Ray Bonn:

The Aberrations of Relativity   by Raymond F. Bonn

This is a collection of articles that place emphasis on the most observable aspects of relative motion. These are the effects of aberration. The author defines what he calls "observational relativity" based exclusively on such directly observable effects. The reader will gain valuable insights into all aspects of relativity using the many informative diagrams and illustrations provided, whether the author's alternative interpretations of Einstein's formulation are accepted or not.

This volume contains 202 pages with many illustrations. It can also be purchased at CreateSpace.

Cosmological Effects of Scattering in the Intergalactic Medium
                    by Raymond F. Bonn

This book provides an explanation of electromagnetic scattering effects that occur in the intergalactic plasma medium, which produce what have been misinterpreted as evolutionary effects. Scattering can account for effects that have been attributed to an origin of the universe in a 'big bang', non-observable 'dark matter', and the mysterious 'vacuum energy' that has been proposed as alternately accelerating and decelerating the expansion of the universe. The scattering explanation propounds a viable stationary state alternative to the standard cosmological model of the universe with predictions that more precisely match observations without the ad hoc assumptions of the standard model.

This volume contains nearly 700 pages with over 200 figures. It can also be purchased at CreateSpace.

www.ingramcontent.com/pod-product-compliance
Lightning Source LLC
Chambersburg PA
CBHW021419170526
45164CB00001B/16